Deep Time

The Journey of a Particle from the Moment of Creation
to the Death of the Universe and Beyond

D1256905

First Edition Design Publishing

Deep Time
Copyright ©2013 David Darling
ISBN 978-1622873-23-4 PRINT
ISBN 978-1622-873-22-7 EBOOK

LCCN 2013939448

May 2013

Published and Distributed by
First Edition Design Publishing, Inc.
P.O. Box 20217, Sarasota, FL 34276-3217
www.firsteditiondesignpublishing.com

Cover Design – Deborah E Gordon

Second Printing

I realize that I am a maverick, for I can settle for nothing less than the whole. It is a feeling deep in my bones and blood. It is the sense of unity of things: man and nature, consciousness and matter, inner and outer, subject and object – the sense that these can be reconciled.

– Renée Weber

The mind of man is capable of anything – because everything is in it, all the past as well as the future.

– Joseph Conrad

We are the music, while the music lasts.

– T. S. Eliot

A Note to the Reader

I have never read a book on cosmology that left me satisfied. Always at the end the big questions remained unanswered: Where did the universe come from? What is the meaning of life and awareness? And so on. *These* are the issues I wanted tackled. But instead I was given a picture, exquisitely detailed in places, yet full of gaping holes and with only a vague suggestion that future science might furnish the missing parts.

So was conceived *Deep Time*. It is not rigorous, not unbiased, not another popular, dispassionate survey of modern cosmology. Yet it does fulfill this commitment – to leave the reader with an unequivocal view – a complete, self-sufficient model – of *why* the universe is, *how* it came about, *where* it is going, and *what* in all of this astounding cosmic drama is the role of mind and man.

Just as the perspective is personal, so is the style. What follows is essentially a prose poem. And within that poem a myth unfolds about the exploits of a subatomic particle as it journeys from the moment of creation to the most distant future. Through the eyes of our hero particle we see the cosmos around unfold and evolve – even as the particle itself develops in a most unexpected way, But more than that, we learn why *this particular* universe came about – had to come about – and to what end it must continue to progress.

One final point; most of the thoughts and theories contained herein surfaced first in the pubs and cafes and other congenial meeting places where great minds are wont to do their best work. I have merely picked and chosen from a galaxy of ideas already expressed elsewhere, tailoring them to suit my need, strengthening ties that to others may seem tenuous, and molding the whole into a single global concept of the cosmos as I see it. For now suspend judgement, open your mind – and enjoy the voyage.

Postscript to the e-book edition AND new print edition

I've made no changes at all to the text for this new e-book edition. Yes, there have been many developments in cosmology and particle physics since *Deep Time* was written about a quarter of a century ago – the discovery of dark energy, for instance. But much of the core science in the book and its most important conclusions have been little affected. And, to be frank, I've grown accustomed to our hero particle over the years, and didn't want to change a story that still, at its heart, feels right to me.

Table of Contents

Chapter 1: Prelude

Whence come I and whither go I? That is the
great unfathomable question.
– Max Planck

Without time, without space. Without matter or energy. This is the beginning of the universe, and there is nothing – not even a point, not even a void.

Out of this nothingness there arises a stir – an eddy, a flicker, a something inconceivably small. And with that something, as part of it, time, space, and other wonders come spontaneously into being. The lid of Pandora's cosmic box has begun to lift and from beneath it issue all the marvels of creation.

Yet by whose hand has that lid been set ajar? And if the answer is "No one's," then how is the magic of genesis performed?

* * *

Countless myths are told of the creation. Myths both ancient and modern, steeped in wonder, each offering its own special window upon the genesis event. From India and China, from the native cultures of Africa and Australia and North America they come. Summoning all manner of gods and heroic creatures to do the seemingly impossible, to bring the world into being. And not just the world but the sun and moon and stars as well, and, in company with these, all of space and time.

And now these older tales are joined by fresh myths born not of faith, not of archaic wisdom unchallenged, but of science, yet no less strange for all that. Gone may be the gods – gone, at least, is their essential presence at each stage in the shaping of what is real. Now nature alone is seen as potent enough, creative enough, to draw itself into existence.

In the beginning, so these new myths of science would tell us, there was nothing. Absolutely nothing. Not matter, or energy, or space, or time. Then came a tiny hiccup, a trivial fluctuation that transformed nothingness into something. Perhaps, our myth would have us believe, the primordial nothingness was unstable.

Remarkable. The universe born out of nothing, of its own accord. But no – not entirely nothing. For if time itself had its origin with some capricious inaugural event, then how did that event manage to occur at all? How could the act of creation begin outside of time?

Unless the rule book of nature was written prior to genesis, how could a state of unbeing know that it had to change? Is nothingness not a much simpler condition, therefore one more likely to prevail, than that of a universe teeming with exotic forms of matter and energy? To all appearances the absence of anything could hardly be more perfect. Why should it sully itself with the seed of stars and stargazers?

This is the central dilemma of genesis – and it afflicts all cosmologies, ancient and new. Wherever the universe came from, before it could emerge there had to be guiding principles, preexisting natural laws. But where did those laws come from? And, in any case, how can a law exist disembodied and outside of time?

Perhaps "before" the laws of physics came the laws of logic, so that the physical laws chosen were the only ones, in combination, that proved logically consistent. But who said the cosmos had to be logical? And whence did the rules of logic appear?

* * *

Time is a marvelous trickster. But one of the greatest hoaxes it perpetrates is to make the creation of the universe seem like the beginning of everything.

Imagine a stream that courses down a tall mountain. At the foot of the mountain, on the banks of the stream, there dwells a tribe. To the people of this isolated commune the stream, with its clear, refreshing water, is all-essential. It is their very lifeblood. And so, because of this, it is also the focal point for the musings of the tribal wise. Where does the stream come from? What is its true beginning? So steep and high is the mountain that none can scale it to seek a definitive answer. And so the wise contrive their theories and spend their days arguing this way and that. It is the god of the mountain, say some, whose tears, shed for the loss of his beloved son, tumble down as the waters of the stream. No, insist others, that is only an admission of ignorance. The stream must somehow issue naturally out of a crack near the mountain's summit. But what happens within the crack – there remains a mystery.

Each day the tribe is blessed with cloudless skies. But almost every night, when the people sleep (and they sleep very soundly), it pours with rain. The rain falls on the mountaintop, collects as a stream, and serves, with each new day, to sustain the tribe and its puzzled priests. Farther down the valley, where these insular folk never venture, the little stream grows to become a river. And, after hundreds of miles, the river reaches the sea, whose water then evaporates to form clouds, which in turn drop rain on the great mountains, to feed the stream that nourishes the tribe. How short-sighted of these primitive folk never to have realized all this!

But then, what of the universe? To the high priests of science, and philosophy, and theology, that, too, is usually regarded as having some special point of origin. And yet is this not just as myopic a view as that held by the sages of our imaginary tribe? The stream, it transpired, had no true source, no real beginning. Might the same not be true of the cosmos?

* * *

There is only one solution to that greatest of all mysteries, the origin of everything. But to understand it requires that we go on a mental journey, perhaps the most daring ever undertaken. It is a voyage into Deep Time, a voyage that begins with genesis and ends in the very remote future of a universe that, quite astonishingly, contrives to become aware of itself.

Chapter 2: Symphony

And God said, Let there be light; and there was light.
– The Book of Genesis

There are no landmarks here. No galaxies, no stars. No elaborations of matter of any kind. Only one second ago the universe itself began in the most titanic of explosions, an explosion in which all the matter and energy there would ever be spontaneously appeared and was borne swiftly out upon the stretching fabric of space. One second old is the universe, and it is filled exclusively with the steaming, fertile brew of creation.

Out of that brew, in due course, will emerge minds fashioned exuberantly of stardust. Presently, these will awaken to their astonishing origins. They will begin to appreciate that they are, quite literally, star children. That they are the universe peering in upon itself. Some of them will even organize patterns of symbols that they might better share such self-reflexive thoughts with their fellow stardust minds. All this is inevitable. So, too, the emergence of sea gulls and spaceships, solar flares and Chevrolets.

But for now there is no such complexity. There is only the raw, untamed potential for it. Potential that, within the one-second-old cosmos, takes the form of a host of loose subatomic particles darting about, colliding, and interacting. And doing so, moreover, not aimlessly, but, rather, in mysterious, blind allegiance to some grand natural code.

Each of these particles eventually will have its own fabulous tale to tell. Of its exploits in the misty netherlands close to genesis. And of its adventures, too, in the greater universe to come, in that vast ocean of time stretching away from creation's shores to those of the most distant future.

Here, now, is one such story, part truth, part fable. A natural history of the universe as glimpsed through the eyes of a single, mythical subatomic particle.

* * *

That particle, the paladin of this tale, chances to be a proton. And as such, as luck would have it, it ranks among the heartier pieces of matter in the infant cosmos. Barely a trillion of its kind would it take, arrayed side by side, to span the dot over this "i"; a mere thousand times that number would suffice to tip the scales against a modest-sized virus.

So comparatively huge is our proton that it seems almost planetlike amid the early universe, a smooth, white, Venus-in-miniature careering through space, with other such particle worlds dimly visible in the distance. Of these some are protons, indistinguishable from our own; the rest, neutrons – similar in size, hued in battleship gray.

Still far off is the age when protons and neutrons will be able to forge a lasting alliance, to come together in tight, permanent clusters. Yet when that does happen these two particle types will emerge as crucial building blocks. They will emerge as building blocks for the nuclei of stable atoms, and so, as the substrata for more mature forms of matter, even for consciousness itself.

Nothing else seems to occupy the newborn universe. Nothing, that is, save for a dazzling, all-pervasive glow. And yet that glow conceals an inner quality. There is a fine texture to it, a shifting, sparkling texture. Indeed, far from being continuous and bland, this iridescent seas swarms with a myriad of lowlier particles, a billion of them for every proton and neutron.

Suddenly, without warning, one of these lesser specks of matter streaks directly across the path of our advancing proton, scarcely avoiding an impact. Like a coal-black meteor it seemed in comparison with the proton's bulk. But there could be no doubting its true identity. This was an electron, a member of the third species of particle that, together with protons and neutrons, are needed for the fabrication of atoms. Its negative electric charge exactly balances the positive charge of the proton. In mass, however, the electron is a dwarf, over 1,800 times lighter than either the proton or neutron. And in size – even a meteor is too lofty a description. For the extraordinary fact is, the

6

electron seems to have no size at all, to be simply a dimensionless point.

In all directions the electrons abound, over 100 million of them to our single proton. And yet they are not the only species of swarming horde. There is also the positron. And here is a very strange thing. For the positron seems to be the complete antithesis of the electron, a sort of mirror image – white meteor to the electron's black.

Nor is the electron unique in having a counter-self. To every fundamental particle in nature there is a corresponding "antiparticle." An antiproton, an antineutron. And so on. Always, particles and their antiparticles are equal in mass. But other of their properties, such as electric charge, are reversed. So, for instance, the positron bears a charge that is as much positive as the electron's is negative.

Most surprising of all is what happens when particle and antiparticle meet. For then they completely annihilate one another, freeing the energy previously pent up inside them as a pair of entirely new particles – particles known as photons.

Everywhere now the crackling of electrons annihilating with positrons becomes apparent, like the noise of some overworked Geiger counter. And yet that mutual act of destruction also prompts a question. For why do so many electrons and positrons remain? It seems that, despite their efforts to extinguish one another, their numbers remain more or less the same.

Re-enter those mysterious photons, products of the electron-positron clash. These make up a third superpopulous particle tribe, as numerous even as the electrons and positrons. But, at the same time, they are a very different breed. For photons, it turns out, rather unexpectedly, are particles of light.

***

Rainbows. A star in Orion. The Mona Lisa's smile. Light may be all of these – or at least the bearer of information about them. And yet – a stream of tiny particles? So, at the subatomic level, may it behave.

A light particle, a photon, is not a material thing. Its substance is not of matter but of pure energy. And the fact that it vaunts energy alone enables it to do what no material particle in nature can – to travel at the ultimate cosmic speed (186,282 miles per second in empty space), but it can travel neither slower nor faster than that. To impede a photon, to trap and examine it, is to destroy it and so free its energy for some new purpose.

7

Most unusual is what may happen when two photons collide. If, together, the photons have enough energy, they may give rise to a fresh pair of particles, one particle plus its antiparticle. What is more, the offspring of this "pair-creation" process may be particles with mass. Mass from energy. Strange. Strange, at least, it seems to dwellers of a macroscopic world. That matter and energy should be two sides of the same coin. That matter is mere patterned energy; energy, the formless clay out of which material things arise.

Now the reason the annihilating electrons and positrons remain so numerous becomes clear. Directly above our proton, two photons – glittering Fourth of July rockets – converge, collide. And abruptly vanish. In their stead materializes an electron-positron pair! And it is the same throughout the rest of this young cosmos. The rejuvenation of the warring electron and positron hosts is everywhere taking place. As fast as the antagonists are slaying one another, new recruits are being introduced to the fray through the efforts of the photons. Which is to say, the electrons and positrons and photons are in dynamic equilibrium so that the number of each, one second after genesis, is maintained approximately the same.

But what of the protons and neutrons? Why are they not being created, along with their antiparticles, from photon collision?

***

Like all newborns, the cosmos is small. Not nearly so small now, perhaps, as it was but a heartbeat ago. Yet very tiny still compared with what it will become.

Imagine all the substance of 100 billion galaxies drawn into a bubble of space only 200,000 miles across – less than the distance from the Earth to the Moon. Such is the condition of the one-second-old universe.

Take a bicycle pump. Pump it several times, and the air inside becomes warm. Compression causes heat.

Take the universe. Squeeze its entire contents into a ball 200,000 miles across. And the cosmic matter becomes hot. Searingly hot. One second after genesis the universe swelters at ten billion degrees centigrade. By comparison, ground zero of a nuclear blast would seem the very essence of tranquility.

Ten billion degrees: At that temperature the energy of a typical photon is more than enough to provide the stationary mass of an electron or positron. Thus from two colliding photons may come an electron-positron pair.

But recall: The electron is an extremely light particle. A proton or neutron is over 1,800 times heavier. So a proton-antiproton pair, for instance, could come only from the union of two photons with at least 1,800 times the threshold energy needed to make an electron and positron. Not even a cosmic oven running at ten billion degrees can cook up such super energetic particles of light. But earlier in the history of the universe? Then might the temperature have been higher still, high enough perhaps for much more massive particles to spring forth from the photons' pas de deux.

As it is our proton moves more securely now. No antiprotons stalk in the particle undergrowth around, threatening it with instant oblivion. All of the antiprotons have gone. So, too, the antineutrons. Presumably, they became extinct through pair annihilation with ordinary protons and neutrons in some earlier age. And that raises an enormous question. For if all the antiprotons and antineutrons have disappeared, then why not all of the protons and neutrons as well? For some reason there must have been an excess of the latter. And that excess is crucial. Because in time it will form the basis for the development of more ornate patterns of matter – like the author and the reader.

With dark mysteries such as these does genesis beckon. So compellingly that to ignore its siren call now and press on into the future – that would be to forgo much, perhaps the very secret of existence itself. The cosmic clock must be thrown into reverse.

And so it is done. Back our proton begins to slide, back down the path along which it came, to descend once more through those micro strata of time that separate it from the origin of everything.

***

Presently that close encounter with the electron-meteor, already described, takes place again, but now in reverse. The electron flashes by, this time in the wake of our retreating proton.

Another electron spears past, perilously near. Followed by another, and another. And then the inevitable: Squarely on, our proton is struck by one of these speeding mavericks. For an instant it seems as if the luckless electron may simply be engulfed, as a fist-sized meteor would be by the Earth, without effect. But then, rapidly, from where the smaller particle buried itself into the larger one's surface, a gray shadow begins to spread outward. Soon that grayness has enveloped the whole of our proton. And our proton is no more. It has been transformed – into a neutron! (In forward time, of course, this would be seen as a neutron decaying into a proton plus an electron.)

Here, then, is another kind of particle reaction common with the early universe. Already, photons in collision have been seen to give rise to electron-positron pairs; electrons and positrons in annihilation to yield photons. Now, it transpires, an electron may merge with a proton to spawn a neutron.

But not simply a neutron. When the electron struck, a second particle, a wraithlike thing, sprang from the midst of the reaction, only to hasten away at vast speed. It was a neutrino, one of nature's exiles, bizarre and aloof.

Billions of neutrinos from the depths of space have passed unmolested through the reader's brain in the time it has taken to assimilate this wild thought. Some came directly from the one-second-old universe itself. Most, by now, have also navigated clear through the body of the Earth as if it were a vacuous ball. Neutrinos move at, or very close to, the speed of light. They are either massless or the possessors of a mass so small that it has not hitherto been measured. And hardly ever do they court the attentions of ordinary matter. A light-year of lead would pose little threat of absorption to one.

But in the subsecond cosmos the state of matter is far from ordinary. The particle soup here is very, very thick indeed. So thick, in fact, that even a spectral neutrino may not move through it far without the risk of interaction.

Just as an electron and a proton may combine to produce a neutron and a neutrino, so, under the extreme conditions of the early universe, the reverse reaction can as easily take place. Within the first second of creation, neutrons and neutrinos find it a simple matter to unite and produce protons and electrons. In complementary style a neutron and a positron may collide to yield a proton and an antineutrino. And to complete the round: From colliding protons and antineutrinos may come neutrons and positrons.

All of which suggests, at first, unbridled confusion. In this primal subatomic riot, incited by ultrahigh densities and temperatures, the various particle species seem capable of interacting at will, randomly, in any fashion. Protons, neutrons, electrons, positrons, photons, neutrinos, and antineutrinos– the main particle types in the cosmos at this stage – all are implicated. And the result: apparent chaos, apparent anarchy.

And yet, in truth, there is an underlying order, well concealed perhaps, to what goes on here. The particles cannot combine ad libitum. Nor are the results of their collision altogether arbitrary.

***

The universe is like a game of chess. It has rules. And those rules govern the actions and behavior of everything there is, including subatomic particles, just as the rules of chess dictate how chessmen may move.

For example, there are conservation rules. Certain quantities, nature demands, must pass through a particle reaction unaltered. Among these – energy. The summed energy (including energy in the guise of mass) of the particles entering a reaction must exactly balance that of the particle products. Energy, overall, can neither be created, nor destroyed. Only can it be reforged, recast.

And, in like manner, nature insists always that electric charge be conserved. What charge goes in must also come out. As when our proton mutated into a neutron. Proton and electron (total charge: +1 + (-1) = 0) gave rise to neutron and neutrino (total charge: 0 + 0 = 0). Had our proton, on the other hand, met with a positron, it could not have produced, say, a neutron and an electron. For then, in flagrant violation of charge conservation, the total charge would have had to change from +2 to -1.

Thus nature sets strict limits on the possible variety of ways that basic particles can mix and match. And yet not only through the conservation of energy and charge does it do this. There are other quantities, too, that it decrees be left unaltered, quantities wholly unfamiliar at the macroscopic level, and so apparently arbitrary that they suggest a much deeper significance to nature's rules. They suggest that not only do these rules control how matter and energy behave but that they also determine exactly the basic forms that matter and energy can take.

Which returns us to our chess analogy. For is this really not also the case with chess? That the rules of the game govern not simply the movements of some pre-existing chess pieces but that those rules, in some sense, actually create the pieces. Forget, for a moment, the superficial appearance of a bishop or a rook or a pawn. The way it is carved, whether of ivory or wood, even its physical presence on a physical board – these are not what fundamentally matter (chess, after all, can be played in the head). It is the definition of what a piece can do, how it may move, how it may interact with other pieces, that is its true and central essence. And that definition follows directly from the game's precepts.

So it is with the universe. The laws of nature lay down precisely, uniquely, the fundamental modes in which matter and energy occur. They define, in other words, the scope and properties of subatomic particles. Moreover, they bring into being the intangible essence, the potentiality, of those particles. And yet here is a critical point: Why those specifically?

Just as in chess there is a small repertoire of pieces – a rook, a queen, a knight, and so on – so in nature there is only a limited number of different particles. There is an electron, a proton, a neutrino. But, for example, there is no particle with ten times the mass of the electron and twice its charge. Nowhere does nature's design provide for such a creature, just as surely as chess makes no allowance for a piece that may leap over three others in a single move.

The cosmic rule book defines just certain types of particle. But not only that. It defines them in such a way that they seem to fall quite neatly into family groups. One such family, the baryons, or "heavyweight" particles, includes the proton, the neutron, and their antiparticles. A second clan, the leptons, or "lightweight" particles, includes the electrons, the neutrinos, and their antiparticles.

Now those additional conservation rules, alluded to earlier, make their appearance. For the universe, it turns out, is meticulous about keeping the separate populations of leptons and baryons at a constant level. Always, in any particle reaction, the total number of baryons less antibaryons and the total number of leptons less antileptons remains the same. And that is extraordinary, for these quantities – the so-called baryon number and lepton number – play absolutely no other part in the physics of the cosmos. Whereas charge and energy, for instance, as well as being conserved quantities, exert obvious dynamic effects, the baryon and lepton numbers seem merely to be a cosmic accounting device.

Why should the universe be this way, why so incredibly contrived? Why should it follow one particular set of rules – one seemingly so arbitrary rather than any other? Who or what made this special selection?

We are like the priests of that fictitious tribe, puzzling over the origin of their mountain stream. Yet we are also bolder than they. For now we are heading up the mountain, swimming back against time's current, moving ever closer (so we guess) to the ultimate source of the cosmic stream on the topmost peak.

***

Our proton has reached to within just one tenth of a second of Time Zero. And now space seems more confined, more congested than ever. The temperature has soared to a tropical 30 billion degrees; the density has risen to over 10 billion trillion trillion times that of lead.

And yet, spectacular though these changes may be, they are quantitative only. True, the particle soup has grown hotter, thicker, even than before. But in content or structure it has not much altered. And that is perhaps surprising, even disappointing. For now the time for such significant change appears fast to be running out. Only one tenth of a second remains before the supposed wellspring of creation itself comes into view. And what is that mere finger-snap of time compared with the billions of years available for cosmic evolution to come? Is genesis about to be exposed as a non-event, as the greatest anti-climax of them all?

Not quite. Time, that master magician, is up to its tricks again. For now it would have us believe that it is a simple thing, a linear thing, always to be considered in equal intervals, each of the same importance (an illusion our clocks serve to strengthen). And yet, of course, that is nonsense. What matters, what is crucial, is not the size of a time interval but what happens within it. During the first one tenth of a second more took place, very much more, than in any similar-length period that followed. The universe underwent more transformation, experienced a greater richness of events, in that first split second that in all the billions of years of star-making and galaxy-building to come.

But how to grasp such a concept? How to become aware of the immense potential for change within even the most slender time slices close to genesis?

First, we must abandon the usual linear way of depicting time. In its place set up an exponential scale in which successive, equally-spaced divisions signify not some fixed increment of time but rather a tenfold leap in its value from the previous division. Bring in, too, a new style of notation. So that, for example, the time of one second after the start of the universe is shown as 100 s, on the exponential scale.

Immediately before this comes the division marked 10^{-1} s, or one tenth of a second; immediately after, the division labeled 10^1 s, or ten seconds. So organized, the axis of time accords the same priority to the period between one tenth of a second and one second as it does to that between one second and ten seconds.

Continuing the exponential scale into the past, further milestones are placed at 10^{-2} s (one hundredth of a second), 10^{-3} s (one thousandth of a second), and 10^{-4} (one ten-thousandth of a second). And now the wealth of opportunity for cosmic development within the first few moments of creation starts to become clear. For on this new scale the "gap" between one ten-thousandth of a second and one thousandth of a second is as broad as that, say, between 10^3 (a thousand) seconds and 10^4 (ten thousand) seconds. The possibility, at least, emerges that what previously may have been dismissed as trifling instants of time near to genesis could, in fact, rival in importance much longer intervals (as measured linearly) that occurred later on. Taken to an extreme, the potential for significant change within the universe may have been as great, for instance, between 10^{-36} and 10^{-35} of a second (a gap of only nine trillion trillion trillionths of a second) as it was between 10^{18} and 10^{19} seconds – a span of over three billion years!

Of course just placing the whole cosmic timescape within a fresh frame in no way alters time's basic nature. That is not the point of the new scale. The point is to highlight an aspect of time that already exists but is easy to overlook. Namely, that time can accommodate a vast number, an unlimited number, of events of critical significance even within its smallest recesses. That, in fact, the universe can evolve out of all recognition within one of time's most fleeting moments – especially when that moment happens to be among the very first.

Think of a human analogy. Think of someone who, as an adult, may take months or years to master an important new skill. Who would guess that within a similar time space, as a young child, that individual had discovered how to breathe, eat, accurately recognize faces and objects, crawl, walk, draw, even talk – skills of astonishing complexity and subtlety.

The universe, too, as a child, learned at a remarkable rate. It learned faster when young because, like a human child, it was much smaller, and all of its internal processes ran correspondingly swifter. The closer the time to the moment of its birth, the more frenetic was its activity and its rate of development. Brief though one tenth of a second

14

may seem in later context, to the juvenescent cosmos it must have appeared as endless, as eventful, as any hot summer in childhood.

***

So now, imagine: Ahead, in the midst of the early subatomic melee, our particle. Below, extending in a thin luminous band across the base of our field of view, a scale like that of a clear tape measure. Time values, marked exponentially on the scale, drift slowly, steadily, past a fixed pointer that reveals the current time reading. Its value now: 10^{-2} s AG, one hundredth of a second After Genesis.

And increasingly our particle hero seems to be suffering from an identity crisis. Under heavier and heavier bombardment by electrons, positrons, neutrinos, and antineutrinos, it flits ever more rapidly back and forth between neutron and proton states, from gray to white and back again. So much so that the generic label "nucleon," embracing both proton and neutron, from this point on will serve it more aptly.

All around, the dawn particle sea roisters. Amid seething waters of 100 billion degrees our nucleon now moves, buffeted and tossed by the currents of a fluid inconceivably more dense than any ocean on Earth.

Yet still the blend of particles is more or less the same as before. Liberal amounts of the finer ingredients, electrons, positrons, photons, neutrinos, and antineutrinos, remain delicately, but essentially, spiced with the much coarser nucleons.

Onward our particle flies. Beyond 10^{-2} s AG, beyond 10^{-3} s, ever on, in a madcap dash to keep its appointment with genesis.

Now the cosmic chronometer reads just 10^{-4} s AG, one tenthousandth of a second after the start of it all. And at last, it seems, the promise of change hangs heavy in the air. Out of a nearby collision of two photons, a proton-antiproton pair appears. A moment later, in the middle distance, a neutron and antineutron emerge in similar style. More and more the process is repeated, as the universe surges past the critical temperature, about ten trillion degrees, at which nucleons and antinucleons form.

But remember: Throughout all this the cosmic clock is running backward. So that, in fact, what is happening now is the final destruction of the antiprotons and antineutrons seen in time-reverse order.

Further our particle moves into the past. Further the universal temperature climbs, well beyond that of the nucleon threshold, so that greater and greater are the numbers of protons, neutrons, and their antiparticles being spawned in the space around.

By 10^{-6} s AG the nucleon-antinucleon population has risen outrageously, 100 millionfold. Risen to be as high as that of the leptons and photons, for by this time all of these particles are being created and destroyed continuously, and with equal ease, during subatomic collisions.

Now that riddle of the antinucleons' disappearance, hinted at earlier, stands firmly astride our path, demanding our attention. For these are the stark simple facts:

At 10^{-6} s AG the numbers of nucleons, antinucleons, and photons are, to all appearances, equal. Yet by 10^{-4} s AG only one nucleon survives for every 100 million photons, and there are no antinucleons left at all. Since the sole way nucleons and antinucleons can be destroyed is by annihilating with each other on a one-to-one basis, we are left with a remarkable, almost unbelievable, conclusion. Namely, the original populations of nucleons and antinucleons were not exactly the same. There is an excess, in the one-microsecond-old universe, of roughly one nucleon for every 100 million nucleon-antinucleon pairs. And tiny though that excess may seem, it is both intriguing and momentous. Intriguing because it proves beyond all doubt that the rules of nature have a peculiar built-in bias for matter over antimatter. And momentous, as already suggested, because that relatively small band of nucleonic survivors will become the basis for all the important structures in the universe to come.

Why should the cosmos not be completely even-handed in its treatment of matter and antimatter? The origin of asymmetry, any sort of asymmetry, seems as incredible as that of the universe itself. Who told nature it had to be lopsided?

But even as the question is asked, our attention is diverted. For now, just as the time gauge slips past 10^{-6} s AG, another tremendous change takes place. Our particle world, our seemingly secure rock in this unsafe cosmos, abruptly breaks up! In a flash its large gray-white spheroid form is gone, shattered. But shattered not into various odd, sizable chunks, like a real planet that had disintegrated, but into three of the tiniest specks imaginable. No bigger than the pointilistic electrons are these. And their name – quarks.

**

Faster than we can follow, two of the quarks born of our nucleon's demise escape, losing themselves in the particle maze around. Only one remains on center stage to become our new hero, or rather our old hero in its new and lessened form.

And the other nucleons in space? They, too, it seems, are bursting spectacularly apart, showering space with their component quarks and antiquarks. Neutrons and protons were not, after all, truly fundamental particles. Each was a bound state of three much smaller flecks of matter, a bag of quark triplets.

But how did the quarks come to be in such bags? The answer to that has to do with the nature of quarks and, in particular, with the nature of the force that operates between them, which is very odd indeed. Quarks attract one another with a force that actually becomes stronger the farther the quarks are apart, as if each were joined to its neighbors by means of an unbreakable elastic thread. Closely packed, quarks behave as if they were free – almost as if they were content that others of their kind were nearby. Yet if any attempt is made to separate them, the quarks immediately counter this by pulling each other more and more powerfully together. So intense is their social urge that quarks simply cannot be made to exist in isolation.

Now, apparently, at about 10^{-6} s AG, a watershed has been reached. No longer can the quarks remain as a single, closely knit, cosmos-wide community. The density of matter has already become too low for that. Driven by their irrepressible need to remain always very near to at least some of their kind, the quarks separate into groups of three. Always three – three quarks or three antiquarks – an arrangement that seems to satisfy well their communal desire. Now, after the first one millionth of a second or so, and for all the rest of time, the universe will be devoid of "naked" quarks. Try to pluck one of its nucleonic bag and the attempt will be foiled. For each quark is like the end of the elastic thread of force that joins it to its neighbor. Can one end of a thread be made to exist on its own? No: Because in the attempt to isolate one end, the original thread is snapped in two, each new piece with its own pair of ends!

But now already our own quark has moved deeper into the past, back to a time when many of its kind still roamed free. Ever smaller dwindles the number of nascent nucleons and antinucleons as their contents break free to join the primal ocean of quarks all around.

A mere 10^{-8} s AG reads the time gauge. And now all protons and neutrons and their antiparticles are gone, leaving the universe replete with an inconceivably hot and dense consommé of leptons, photons, and quarks. Not one of these particles boasts any extension in space. Not one has the solidity or the apparent propensity for organization that matter in the future universe will have. And yet each of these spots of next-to-nothingness seems to "know" exactly what it is; even stranger, what its rules of engagement are with other particles. And that seems most peculiar. It seems to bring us close, in fact, to the very heart of what reality is. For now we are dealing not with stars, or rhinos, or political intrigue, or with any such higher level of complexity. One hundred millionth of a second after the universe began, we stand in the presence of near-ultimate simplicity.

***

Ironic that something so simple should prove so enigmatic, so hard to understand. We thought we knew matter, knew it intuitively at least. Just as we believed we had an innate feel for those other rudimentaries of nature, energy, space, and time.

Matter was? Tangibility, substance. It was the ground beneath your feet, this book in your hands. And yet all that is now seen to involve a much more abstruse concept. What we experience and label as "matter" in our macroscopic world is actually some vastly elaborate and entangled hierarchy, as far removed from the basic underlying essence of matter as a Wagnerian opera is from each of its component notes.

Matter – raw, unembellished matter, the protocosmos reveals – consists of quarks and leptons. And nothing more. All there ever will be in the universe must be fashioned from these insubstantial parts. (The photons, too, are still present in huge numbers but are constituents of energy.)

But a quark and a lepton have no size. And yet, somehow, they must contrive to make "objects" that we perceive to have physical extension. Objects like you and me, and galaxies.

And there is more to this mystery of matter. Because, of course, the quarks and leptons of the early universe are not individually aware of the task ahead of them – that they have to come together to make neutrons and protons, atoms and molecules, stars and galaxies, and, eventually, cosmologists. Yet at 10^{-8} s AG they are also everything,

18

materially, that there is. And from them, indeed, progressively more interesting creations will come.

Blind and primitive the dawn quarks and leptons may be. Yet they are destined for greater things. Though their febrile dance has the appearance of being chaotic, out of that chaos, we know, will emerge order. It is as if they carry within them a code, like a colossal genetic code, for building the universe to come.

***

But building presupposes the ability to join parts together. So that not only must there be fundamental particles, but there must also be fundamental forces by which to bind them.

In chess, too, forces are at work. A chess piece "interacts" with another when it captures it. And there are several ways this may happen – several types of chess force. Certain of the pieces, for example, can strike from one or more squares away, an ability to capture we might describe as "long range." On the other hand, a piece such as the king, say, is limited to attacking its victims solely from an adjacent square, so that its influence is clearly more "short range." We could even go on to identify subcategories: a second kind of short-range force, for example, as exhibited by the pawn during en passant capture. The essential point, though, is this: Different chess pieces exert different types of force, which, in turn, are defined by the rules of the game.

And so it is with the cosmic chess game. In this case the governing laws, it turns out, prescribe four forces. Not eight, or twenty-two, or any other number, but precisely four: a quartet of natural forces that operates within the universe, one hundred millionth of a second after its formation, as it will for all the rest of time.

Of these cosmic forces two are long range, capable of acting over any distance. And for this reason they are the ones well known to us – gravity and electromagnetism. Gravity is the force that operates between all things in the universe with mass. Gravity, the planet builder, the apple dropper, the star crusher. But gravity, too, the incredibly feeble: Compare it with electromagnetism, that force which all electrically charged particles exert on one another. A pair of protons push apart more powerfully through electromagnetism than they pull together through gravitation by a factor of 100 million trillion trillion trillion! Only because, overall, space is electrically neutral does gravity come to dominate over cosmic scales.

At very much shorter range, when subatomic particles are in near contact, two other, totally unfamiliar forces may play their hand. These are the strong and the weak force: the strong, the strongest force of them all; the weak, ten trillion times feebler than electromagnetism, yet still awesomely large by gravity's standard.

Just as in chess, where not every piece may exert all of the forces that the game allows, so also in nature. An electron, for instance, can take part in interactions through gravity, electromagnetism, and the weak force, yet neither can it exert nor "feel" the strong force. The last is the reserve of the quarks.

So perhaps, indeed, the universe is merely a gigantic board game. It has pieces (particles), a board (space), and, apparently, its own set of rules (defining the particles and the ways they can interact). Presumably, it began with its pieces in some certain arrangement, the opening moves were made, and now the great cosmic game is underway.

Yet the analogy is far from perfect. And that for a very good reason: We can look beyond an actual chessboard to see the human hand that moves the pieces. We know why chess is the way it is, why it has its own ad hoc set of rules. We know because we devised it. We are the gods of chess who look down on it and control it and understand everything about it from our "meta-chess" world. But how can we say that of the entire universe? No convenient metacosmos exists to afford us a true god's-eye view (and if it did, we should need a metametacosmos from which to look down and understand that). Nor can we glimpse any obvious outside agency guiding each particle in space moment by moment. The universe seems to run itself, to make its own game moves from within.

As for the setting up of the cosmic board, and the devising of the cosmic game rules, here is the greatest puzzle of them all. For, like chess, the universe seems to have one set of rules, one set of pieces, one board. It exists in a certain way. One hundred millionth of a second after its birth, it contains just certain types of particle, able to interact through just a handful of specific forces. But why?

In the case of chess such arbitrariness is no mystery. We know chess to be a capricious product of the mind. We accept that its human inventors could as readily have compiled a quite different set of rules (some did and came up with checkers and Go).

But the contrived nature of the universe is not so easily understood. Who or what devised the cosmic rule book, laid down the great board of space, and set up the particle pieces? Why those rules, that board, those pieces? Why not, ten nanoseconds after the game began, a cosmos with seven different forces? Or none at all? Why not with a cosmic collection of pieces wholly unlike the quarks or leptons? And where did all this fantastic ensemble of matter and energy come from anyway?

Ever the wide-eyed child within us asks such questions, now as it did years ago: How did all this come about? What made the world? And why this world?

Ever the wise old scientist within us strives to reply, to reconcile and satisfy:

"I know the universe seems contrived. But since we're here discussing it all, it must exist in some state. And whatever state that happens to be is bound to appear special!

"At the same time, I understand your concern. 'Why,' you want to know, 'did the universe evolve in this way and not some other?' Just bear with me if you will. Because I believe I can fully answer that. My plan is to study the universe as it is now, just as I might examine the state of a chess game sometime after it has started. Then, once I have figured out all of the rules, I shall work backward, move by move, to deduce the original state of the game. Give me another ten, or fifty, or a hundred years and I'll have all the essential detail of the early universe worked out. I've every confidence I can push my way back to the moment of genesis itself. You wait and see."

***

Now our time indicator shows 10^{-9} s AG. And the entire cosmos has shrunk so much that it would fit comfortably inside the globe of the future sun – a tiny womb wherein the most primitive forms of matter and energy incubate.

Deeper, ever deeper, our particle descends into the eye of the swirling vortex of counter clock time. To within one ten billionth of a second of creation it moves, while around it the temperature soars to ten thousand trillion degrees. And now another transition begins, more remarkable, more profound even than the freeing of the quarks from their nucleonic jails. Once again it is as if the universe had tripped some hidden mechanism, a mechanism seemingly precontrived and awaiting only the right time and conditions to set it off.

Four cosmic forces there had been: electromagnetism, gravity, the strong force, and the weak. Four basic ways in which the humblest particles in nature could interact. But now, at 10^{-10} s AG, an extraordinary change is starting to take place. Electromagnetism and the weak force, previously so disparate in character, are fast growing more and more alike. As the temperature rises further and time edges back to just one trillionth of a second AG, so these two forces gradually meld and become one. The product of their improbable union: the electroweak force.

(Again, it must be remembered, this is in time-reverse order, from future to past. So that in truth the universe started out, prior to 10^{-12} s AG, with three distinct forces. One of these, the electroweak, then began to fork. And thereafter its twin offshoots went their separate ways, never to reunite.)

Even so, the primal kinship of electromagnetism and the weak force would not be lost, just as in some future age man and the apes would diverge from common stock and yet still retain the evidence of their shared ancestry. That affinity, between life's evolution and the development of the early universe, runs deep. Both are marked unmistakably by the growth of complexity. Four forces from three. Compound particles from simple. Man and ape from a single, prehominoid ancestor. Revealed here is a common omnipresent urge of the cosmos to unfold, to progress, by its own efforts, toward some unseen final goal. This same urge it was that prompted both the splitting of the electroweak force and, billions of years later, the division of the line that led to man.

Yet how far back can this process of complication be traced? Are the three forces and the handful of elementary particle types in the universe at 10^{-12} s AG themselves the products of an even earlier chain of evolution? Or do they represent nature's ultimate, irreducible state?

Onward our quark explorer presses in quest of Time Zero, its energy increasing without bound as it penetrates farther into the past. Beyond 10^{-13}, 10^{-14}, 10^{-15} s AG, it goes, while around it the temperature and density of cosmic matter rise and rise to new inconceivable heights. Trillions of encounters it has, but always these involve the same triad of forces and the same basic set of rules.

Now, at 10^{-20} s AG on the exponential time scale, our particle has crossed more orders of magnitude than those that separate humans from the one-second-old universe. Even so, its immense, regressive journey

22

into Deep Time has barely begun. Still, at 10^{-25} s AG, there is no sign of any further qualitative change. The separation of electromagnetism from the weak force now lies at an awesomely remote point in the future. Was it, after all, the first crucial cosmic development, the first-ever step toward great complexity?

And now, farther back our hero flies, to 10^{-35} s AG. Suddenly, with the universe ten trillion trillion trillionths of a second old, the question is answered. Just as before the electromagnetic and weak forces conjoined, so now, at this much earlier time, the strong force begins to merge with the electroweak. It is a stunning event − a triumphant triple alliance that brings to the universe, at last, an even simpler and more perfect order.

Now there are only two distinct ways in which any piece of cosmic matter may interact. There is gravity. And there is the grand unified force. Embraced by the latter are all of the attributes previously associated with the strong force and the weak and with electromagnetism. It is as if all the pieces on a chessboard − pawns, rooks, and bishops alike − were given the same freedom to move around and coact as a queen. How could any of the pieces then meaningfully be told apart? When the strong force united with the electroweak that same freedom was bestowed upon the quarks and leptons. Suddenly anything a quark could do a lepton could do equally well. A lepton could even turn into a quark, a quark into a lepton.

In this new democratic universe no longer is our hero particle distinguishable from any of the others with which it interacts. The differences between quarks and leptons, so conspicuous before, have vanished. They are revealed as having been mere facades, skin-deep variations, hiding a common core within.

Yet even as the universe assumes a form still simpler than before, the mystery of its origin and unfolding remain. At 10^{-35} s AG all that there is from which to build all there will ever be is this pocket sea of primitive, like particles − the quark-leptons. But what drove these cloned specks of energy to evolve and combine, through many branching levels of complication, into atoms and galaxies and minds? Despite their overt simplicity, the grand-unified particles bear − even at this early stage − the inner potential for what is to follow. Truly they are the spores of the cosmos, the harbingers of complexity to come. But even spores must have an origin, as must the concealed genetic message they carry.

***

So, to the final stage of our assault on genesis. And now we must venture on alone, forsaking the company of the particle with which we have journeyed thus far. Already it is moving away from us, our last direct link with the cosmos we knew. For a while we watch it still, colliding, merging with its protagonists, then reasserting its individuality once more. And yet no longer can we really be sure whether this is "our" particle or some other that is identical to it in every sense.

So at last we turn away. And prepare our imagination – to accept a universe that has shriveled to the size of a pea. That has become so heated – to 100 trillion trillion degrees – that every particle within it has the energy of a charging bull. And that, like an equatorial dusk, is rapidly about to ...

Darken.

What a stunning, almost biblical, genesis it would have made – this, the brief, spectacular era of the creation of the particles. Here, at 10^{-35} s AG, or a little before, most of the matter in the universe, sprang into being. Photons too – explaining, in time reverse, the sudden dimming. As part of the process by which the strong force uncoupled from the electroweak, by which the grand unification was broken, most of the material that the universe would ever contain showered into existence.

And yet, surprisingly, this is not the true beginning. It may be the moment when the cosmic cast of matter first appeared. But before that there was still the empty, waiting stage – still space, still time, and, in some curious abstract form, still the laws that regulate every physical action, including the birth of matter. So we must search on to find now the source of the backdrop against which the particle actors will play out their drama. And we must search in the dark, groping, shedding what dim light we can by our own mind's eye.

Presently, with the distractions of matter gone, the very geometry of space and time begins to make itself apparent. Subtly woven together, it seems, space and time form a surface upon which all the events of the future universe may take place. And yet this topographic "space-time" is neither static nor necessarily flat like an artist's canvas. It can move – swell or contract – and it may warp and twist, even curve back upon itself to yield a closed shape like the surface of the Earth. But whereas the boundary of a normal sphere has only two dimensions and is curved in a third, the surface of space-time has, from the outset, an unimaginable dimensionality of four (three of space, one of time). And so any warping of it must involve a still more recondite, fifth dimension.

Only through some strangely heightened awareness – an ability to perceive more than three dimensions at once – are we able now to trace the further devolution of cosmic space and time. We "see" in n dimensions, where n is a number indefinitely high. But we are compelled to set down our impressions here, imperfectly, in just three. At 10^{-40} s AG the universe is the surface – so we imagine – of a microscopic ball, still shrinking.

In speculative vision we see the ball-universe from outside, a faintly glowing orb suspended in blackness. The chronometer reading is 10^{-41} s AG. Far smaller than a proton now is the vacant, embryonic cosmos; we, circling around it somehow, occasionally swooping in to view more closely the texture of its space-time skin. Smooth and continuous this had seemed at first. But now, at 10^{-42} s AG, the first signs of a less than perfect complexion are beginning to appear. The surface of the ball-universe, near to, looks finely speckled. And the speckles are dancing, vibrating chaotically like a swarming mass of flies.

So in this myth within a myth is the motile graininess of space-time itself revealed to us. But even at that instant of discovery – our chronometer marking the time at 10^{-43} s AG – we are overwhelmed by a far greater surprise. Rising up to meet the tiny, luminous drop that is our nascent universe is a tall, slender, shimmering spire. And as in bewilderment we look down this, from the tip of its needle like summit, down its sheer sides, to the gentler sloping curves of its base, we see that it is itself merely a trivial and temporary feature thrown up from a vast, turbulent ocean of pandimensional space-time. With sudden, terrible insight we realize that our cosmos is like an insignificant speck of foam that has escaped from the endless, restless sea. Escaped, sealed itself off, grown, nurtured a hundred billion galaxies, and . . .

A speck of foam that is now becoming stretched into a tear-shape, to be re-joined, as we see it in retrospect, with its cosmic mother sea. The spire makes contact, fuses at its tip with the errant droplet, then descends with its long-lost prize back into the froth and spume below.

***

25

Shocked by the suddenly, brutally widened horizons of our vision, we slam shut our cosmic senses. This must be some crazed dream or a fantasy brought on by overzealous imagining. And yet:

"It is neither," urges the familiar voice of the scientist within. "Simply, you have been witness to one of my more engaging reconstructions of genesis. Allow me to explain.

"As you watched our universe approach the actual moment of its birth, the previously hidden, inner structure of space and time came into view. You began to see the elaborate, shifting microarchitecture of space-time. When, finally, the time intervals you could discern were as small as 10^{-43} of a second, the characteristic lengths no greater than 10^{-32} of an inch, the full evanescence of the cosmic fabric became clear. At these scales the very concept of distance and duration loses meaning. Tiny new regions of space and time can spring capriciously into existence or, just as readily, die out.

"One of these regions, by chance, broke free from the great primordial sea of space-time – and, in due course, evolved into the cosmos you know. The instant at which it detached itself from that slender, rising stem – the instant it became an independent 'child universe'– was, if you will, Time Zero, the moment of genesis. It was then that the internal clock of our universe began to tick.

"As for the parent ocean of space and time from which our cosmos arose, I can say little. It was a confused region, devoid of stable or specific dimensionality. Perhaps, in truth, it was less like an ocean, more of a fine dust – a pregeometric dust of points, outside of space and time, that occasionally and locally assembled itself to make the seeds of future universe."

And yet that leaves much unanswered: If this "ocean" or "dust" existed outside of space and time – if it was, in some sense, the basal stuff from which organized space-time could come – how then could it change? How could our universe have been "ejected" like a speck of foam from the precosmic ocean? Or, invoking the other metaphor, how it could it have been "assembled" from a collection of points in the ancestral dust? Whether the progenitor of our universe was like a wind-lashed sea or a swirling, ethereal dust of pregeometric particles, if it existed outside of time, then how could it alter, reconfigure itself, so as to spawn a coherent, self-sufficient universe? Change of any sort, however esoteric, however unimaginable, demands the prior existence of time. But then where did time come from?

26

"Yes, I know there are some small details missing. But, you see, all this is such a new field of research. We've barely scratched the surface of the science of genesis. To have tentatively reached back to within just 10^{-43} of a second of creation, to have begun to address the problem of the origin of space and time itself, is – you must concede – no mean accomplishment."

True. Science has achieved much, and it will go on to achieve much, much more. Yet, nevertheless, it is working back along an exponential scale. And before 10^{-43} s AG came 10^{-44} s AG. And before that 10^{-45} s AG. Maybe when physics is a century older we shall have a rationally arrived-at account of what 10^{-100} s AG was like, or even 10^{-1000} s AG, if such times have any physical meaning. Yet still left unsolved will be the mystery of "what came first?" And, in some ways, that is really the only question that matters.

"Ah! I think you're assuming now that there must have been some initial cause of our universe. But in fact, when we probe closely the world of the very small, we find that there's no such simple chain of cause and effect. Particles of matter and energy and – our recent theories tell us – of space and time, too, behave unpredictably. They can appear and disappear, without prompting, without prior intervention. And if they can do that . . . well, it seems entirely reasonable – even quite likely – that the universe itself may have begun as a chance event.

"Exactly what it 'began from' is still, I confess, a moot point. I conjured up the ideas of a natal ocean or a dust cloud. But of course I don't mean you to take these too literally. At best they are devices to help your mind key upon the extraordinary problem of cosmic origins. As soon as you are able to, and if you so wish, cast the devices aside. Do without the crutch of believing there was 'something' there to begin with, a nebulous, indefinable matrix from part of which our universe took shape.

"Instead of that 'something,' suppose there was, quite simply, nothing. And I do mean, no-thing – neither space nor time. Then, by chance, there was a fluctuation. Maybe there were many of them, but only one persisted. It was a fluctuation – a tiny, random, unmotivated flutter – in which a pattern of points appeared. Those points constituted, by their type and arrangement, a primitive space-time. Our space-time. The arrangement proved stable. And so it endured. Not only that but, in time, it evolved – evolved so that regions of it became intelligent and, eventually, curious about their cosmic origins.

"And there you have it. My greatest, most daring speculation – the universe sprang into being, purely by chance, out of a perfect, spaceless, timeless vacuum."

Creation ex nihilo! The supreme scientific myth . . .

Except that it is fatally flawed.

Space and time never could just "spring into being," because measured against what time order would they do the springing? Detail is absolutely crucial here. There can bo no fudging the issue, no playing with words, no leaps of faith. Exactly how could time (or space, for that matter) begin? Describe the process. Describe the very first step.

And now the scientist within is silent. For the unassailable fact is this: Under no circumstances can there have been a time when time did not exist! Never a state of timelessness.

And what applies to time applies equally well to the laws of the universe. There can never have been a time – a pretemporal state – of lawlessness. Because otherwise by what design or mechanism could the laws have been produced? And so carefully, so uniquely selected?

Science has come far, led us almost to the summit of our cosmic mountain, almost to the source of our universal stream of space and time, of matter and energy. Almost. But in the process of seeking out the headwaters of all creation it has confined itself – perhaps necessarily – to reductionism. And so now the scientist-priests in their search for genesis appear like folk on their hands and knees peering within the tiniest crevices around Time Zero for that legendary inaugural event, the primum mobile. If only they can rid themselves, so they think, of the last vestige of time and space and materiality. If only they can show that, in the beginning, there was simply this most perfect, pristine nothingness. And if only they can demonstrate that from this inconceivable absence of anything the plenum of existence naturally, fortuitously flowed – then they will have succeeded.

Nor should we criticize. For reductionism indeed has borne rich fruits of understanding. This whole epic journey through Deep Time is, in part, a celebration of what science has discovered or conjectured about where the cosmos came from and where it is going.

But to grasp the ultimate truth of genesis – that requires an abrupt change of perspective, a new scientific and philosophical outlook, or maybe a radical expansion of the old. For now we must look not to reductionism but to holism. Not obsessively at one certain point in the universe, but at all the universe, future as well as past. And why? Because, as we are dimly now beginning to see, the creation of the universe was in some curious way bound up intimately with what was to come.

The cosmos was not – never could have been – some accidental flower that grew spontaneously from a nothing seed. Its galaxies and earthly flowers and blossoming, self-aware minds are not mere lucky inventions. The universe is exquisitely designed, intelligently constructed, beautiful beyond belief, a creature spawned out of awesome complexity, not austere simplicity.

And now, suddenly, our task becomes clear. It is to search for the source and meaning of the cosmos not in the extreme past but, on the contrary, in the remote future – at the nether edge of Deep Time.

Chapter 3: Kinds of Flowers

The most beautiful experience we have is the mysterious.
It is the fundamental emotion that stands at the
cradle of true art and true science.
– Albert Einstein

Already fading are those strange, shifting impressions of Time Zero, those half-glimpsed images culled from the scientific lore of genesis. With a single instantaneous leap we have returned to a more familiar place: the one-second-old cosmos, with its thick, pervasive stew of protons, neutrons, electrons, photons, and neutrinos, simmering at ten billion degrees.

From this point a new phase of our adventure begins. Inevitably, mysteriously, the path to the future winds away into the gray, misty distance. And even as we strain to see where it may lead, our particle hero, restored to its protonic form, approaches once more out of the bright chaos around.

***

Urgently, unpredictably, our proton moves, as of old, from one brief encounter to the next. Rebounding from an electron, now from a fellow proton, it seems engaged in an endless, apparently aimless game of subatomic billiards.

Not for the last time do we ponder the prospects for this mad jumble. And yet we recall, too, that the universe has already passed through a bewildering series of transformations – in a single second! So how much more might it not evolve in the billions upon billions of years to come?

Still, factored in must be the rate of cosmic metamorphosis. And that now is noticeably on the wane. The next ten seconds bring no fresh revolution, no great new surprise. At the fifteen-second mark, though the temperature has dropped to a balmy three billion degrees, the particle mix remains the same.

A full minute goes by. And more. So that, half seriously, we begin to doubt whether nature has any creative power left following the first frenetic second.

Two minutes AG. Three minutes. (The doubt grows.) Three minutes 45 seconds AG:

And now, as it has many times before, our proton draws near to another tiny islet of matter – a neutron – collides with it, and scatters. Yet, for a lingering instant, the proton and neutron remain attached. Like two drops of liquid that touch, the nucleons momentarily fuse. Only barely is the vigor of their impact enough to prevent the strong nuclear force – operating between the proton and neutron at close range – from binding the two particles permanently together.

But even as our proton flies off, its independence narrowly preserved, the neutron from which it has just pulled free strikes another proton nearby. And this time there is no subsequent scattering. Instead, the neutron and its new partner remain tightly, securely bound together – as a nucleus of deuterium, or heavy hydrogen (^{2}H).

At last, it seems, the temperature has fallen sufficiently for this new stage of cosmic synthesis to begin. Alongside those most primitive of nuclei, the protons (the nuclei of ordinary hydrogen), small quantities of deuterium start to appear. And not only that. Some deuterium nuclei quickly go on to collide with and absorb an additional neutron and thereby change into tritium (^{3}H), the heaviest form of hydrogen. Others, by chance, acquire an extra proton and so transmute into a lightweight variety of helium, helium-3. In either case the normal mode of helium is but a short step away. Tritium swallows a proton and becomes helium-4 (two protons plus two neutrons); helium-3 swallows a neutron and does the same.

At three and three-quarter minutes AG vast quantities of helium 4 are being produced rapidly all over the universe. And yet therein lies a puzzle. For the fact is that helium-4 is sturdy enough to survive at temperatures of around three billion degrees. In other words, it was sufficiently cool much earlier, at only fifteen seconds AG, for ordinary helium to exist. Why then did it take so long to appear?

The answer lies with the temperature stability of the middlemen: deuterium, tritium, and helium-3 – especially deuterium. Three billion degrees is still far too hot for these weaker-bound nuclei to hold together, so that they are simply blown apart the instant they form. Although the end product, helium-4 is stable at much higher temperatures, its formation is delayed by the more fragile nature of its intermediaries.

Only as the temperature slides down to around 900 million degrees, at about 225 seconds, do deuterium and tritium and helium-3 each manage to cling together long enough for the final jump to helium-4 to occur. And then, quite suddenly, it happens: The universe is 10 percent helium, and the dramatic moment passes at which deuterium finally achieves stability. With the chain reaction process from individual neutrons and protons to helium-4 no longer chocked off at the second level (deuterium), virtually all the remaining free neutrons are gobbled up into helium nuclei.

And our proton? Despite some close shaves, it has retained its liberty throughout this early phase of cosmic nucleosynthesis. Though one in ten of the nuclei around it are now of helium-4, almost all the remainder are free protons like itself. A tiny fraction endure as deuterium and helium-3 (though not as tritium, since this is radioactive and quickly breaks up). And there is a small but dwindling tribe of nomadic neutrons.

Unlike the proton, the neutron cannot live indefinitely on its own. Bound up within a nucleus, it is secure. But alone, unattached, it must, as if it were a live grenade, quickly split apart – another strange idiosyncrasy of nature. Isolate any neutron at random and the chances are fifty-fifty that it will decay – into a proton, an electron, and an antineutrino – within just twelve minutes. Every 100 seconds from now on the remaining population of free neutrons will decline by 10 percent, until the only neutrons remaining will be those enclosed within nuclei.

***

Four minutes after the Big Bang: Blindingly intense radiation bathes every corner of space. The photons swarm, 100 million of them for every proton and neutron. Electrons and their antiparticles, the positrons, continue their inevitable annihilation, until all the positrons have gone and the residual electrons are roughly equal in number to the protons. There are the ghostly neutrinos and antineutrinos. And, at the other extreme of materiality, there is this new, complex thing called helium.

But why should the universe stop at the helium stage? Why not go on immediately to build still more complicated nuclei, perhaps those of carbon, oxygen, silicon, or even iron? The reason is the same as that for which the formation of helium-4 was delayed. Even when the final product was stable, certain intermediate nuclei – vital stepping-stones in the process of nucleosynthesis – were still highly unstable. Deuterium's temperature sensitivity caused the hold-up in helium-4 manufacture. Now, for anything heavier than helium, such as lithium-6, beryllium-9, boron-10, or carbon-12, it is the unstable nuclei with five and eight nuclear particles that are the stumbling block. Only in a very different environment, in the dense, central furnaces of stars-to-come, will nuclei more elaborate than helium be able to take shape.

A half hour slips by. Our proton moves within a cosmos cooled to 300 million degrees – just fifteen times hotter than the core of the future sun. Less often now does it collide with other particles. The average density of matter has dropped to just one tenth that of water. Nucleosynthesis has come to a virtual standstill.

And again there is apparent quiescence. Again, after a sudden, frenzied burst of change and synthesis, the anxious waiting for some new, unknown step toward greater cosmic complexity. Only this time the waiting seems longer – interminably long, even by human standards.

Gone forever is the Golden Age of ultrafast transition, when the character of the whole universe could alter beyond recognition within the smallest fraction of a second, or within a few seconds, or a few minutes.

An hour elapses. A day. A year.

A thousand years!

And, all the while, space relentlessly expands, stretching further the kinetic pattern on its multidimensional surface. The density of cosmic matter, along with its temperature, continues steadily to fall. And yet, apparently, there is no change in matter's quality.

Ten thousand years go by. And even though our proton, like the countless other protons and heavier nuclei around it, often passes close to an electron, it forms no partnership with it. Even though the proton and electron have equal and opposite charge, and are therefore powerfully drawn to one another, they fail to come together in stable alliance. X-ray and ultraviolet photons – bullets of high-grade energy – strafe the fledgling cosmos, instantly stripping away any electrons that dare to enter bound states around a nucleus. Laser-intense, ubiquitous, the young electromagnetic field tears apart anything resembling an atom. And so, for millennia upon millennia, while the universe burns this bright, there is only a writhing, thinning, electrically charged mist – a plasma – of loose nuclei and electrons.

Or so it had seemed. And yet there may be more to this young universe than simply a hot, spreading fog of particles and blazing light. For now the saga of our proton is beginning to take a strange new turn. And the prospect is slowly emerging that there may have been other things born of the Big Bang, bizarre, almost indescribable things, that have surreptitiously found their way into nature.

***

At first it had seemed incredible. But now there can no longer be any doubt. Our proton is being pulled, gradually, irresistibly, over thousands of years, toward ... what? Some unknown attractor. An obscure but tremendously powerful source of gravity. And it is not just our own tiny particle but all of the subatomic matter for light-years around that is streaming in toward this unnamed, previously unsuspected phenomenon. Nor is the mystery attractor unique. There are billions of others of its kind strewn about space, each busily spinning its own cocoon of hot plasma. Evidently, even as the universe continues its headlong outward rush, portions of it are being drawn together locally. Material is steadily accumulating around these newfound objects instead of being diluted more and more by the overall cosmic expansion.

Which seems promising. For while matter holds stubbornly together, there is the chance at least of it forming more complex arrangements. While protons, electrons, and the other shards of genesis remain in reasonably close contact, they can hope to interact more often and combine and so further evolve.

Yet still there are no visible clues as to the nature of our proton's lure. Only vaguely, through enhanced perception, do we begin to sense some of its outlandish properties. At the heart of the influence, it seems certain now, is no ordinary object, no familiar aggregation of matter. On the contrary, whatever is doing the pulling is exotic in the extreme – a loop of ... something ... a light-century long, yet no thicker anywhere than a million trillion trillionth of an inch. A shoelace length of it would outweigh the Himalayas. And yet, because of the terrific tension within it, the creature fiercely writhes, like a monstrous captured eel, at speeds approaching that of light. This is a cosmic string – a long, wriggling, ultra-thin tube of energy trapped since the dawn of time.

It was at that primal moment when matter first appeared, when the strong force first split away from the electroweak, that cosmic string first flickered into reality. The process by which it did so resembled the freezing of a pond or lake. Before 10^{-35} s AG the vacuum of space was smooth, homogeneous, the same at all places and in all directions. Before 10^{-35} s AG, the scientist would say, the universe was highly symmetric. But then, as the strong force peeled away, some of that ancient symmetry was lost, was "broken," forever.

In the same way, liquid water has a high degree of symmetry. It appears the same from whatever point or direction it is seen. Yet, as it turns to ice, water forfeits much of its regularity and simple beauty, retaining only certain preferred axes of symmetry where its molecules have lined up in orderly fashion along crystal planes.

And there is a second source of novel complexity here. For of course an ice film never forms all at once across a stretch of water. Rather, the freezing spreads out from various random spots until the whole surface is covered in patchwork style. Yet since the crystals growing at one center cannot possibly "know" what is happening elsewhere, almost certainly they will have symmetry axes pointing in a different direction from those of crystals of neighboring centers. Where the ice fanning out from two points meet, there will appear an irregularity, a discontinuity. In other words, an ice sheet will contain defects where the direction of symmetry abruptly changes.

And this is exactly what happened when the strong force "froze" out, 100 billion trillion trillionth of a second after the Big Bang. Maybe the physics involved was more esoteric, the properties less commonplace than those of a village pond in winter. But the analogy holds true. A cosmic freezing took place at 10^{-35} s AG. And as a result of this the universe acquired defects – wrinkles or internal boundaries – one manifestation of which is the fantastic object toward which our proton is now speeding.

Little wonder then that cosmic string should seem so extraordinary. The stuff it contains has survived unchanged from the era of the grand unification, when the energy density was inconceivably high, when all of space-time was curled up within a region far smaller than a nucleus. Cosmic string has no right to exist – does not belong – in a universe tens of thousands of years old. Yet here it is, a living fossil, playing a pivotal role in the development of more mature forms of matter.

Without cosmic string, we begin to realize, nothing much could have come of the hot breath of genesis. Simply, it would have been swept away, endlessly dissipated. The self-gravity of the cloud of particles that billowed out of the Big Bang would have been far too feeble by itself ever to cause the cloud to become clumpy. And without the prospect of some local infalls of material – of the condensation of knots and swirls of richer material – what hope could there be for more interesting structures to emerge? Structures like diamonds and oceans. Like elephants and senators. And like those of us who strive to understand how all this came about. The strange truth is surfacing that every marvel of creation owes its existence, ultimately to cosmic string.

But then, how did – and how will – cosmic string evolve? That now becomes a paramount issue. Because cosmic string, it seems, both predated and predetermined the large-scale structure of the universe. Just how did string itself evolve? Moreover, what effect is it having – what sort of material structures is it seeding – in the current universe of our particle hero, 30,000 years after the Big Bang? And what further influence will it bring to bear as time goes on?

***

Return almost again to the beginning of the universe. Revisit those moments just before 10^{-35} s AG when there is as yet no matter, no string. When there is only a vacuum. Not nothingness, but vacuum – a state of minimum energy in the absence of particles. Later, the energy of that vacuum will fall to zero. But in the earliest instants of time, the vacuum is enormously energetic, pregnant with creative potential.

Now the great moment approaches. The cosmic pond is about to freeze over. Within many scattered, patchwork domains the phase transition begins, spreading quickly outward. Inside these domains much of the latent energy of the vacuum is discharged spontaneously as a mass of subatomic particles. Matter, in huge amounts, is born. But along the narrow margins, where neighboring domains meet, fragments of the ancient, high-energy vacuum are preserved. Preserved perhaps in various forms – as null-dimensional points (particles called magnetic monopoles) and as two-dimensional sheets (domain walls) – not only as linear threads. But it is the threads that most interest us. For these wormlike "cosmic strings," we suspect, are destined to lay the foundations of the future universe.

37

On our first journey here we failed even to notice the string, so overshadowed was it by the sudden, spectacular arrival of matter. But now, with our awareness trained upon it, we see the string clearly. It pervades the early universe like a tangled web. A web that has no endpoints, for its individual strands are either infinitely long or fashioned into closed loops.

Wriggling violently because of the tension within them, the strings often cross themselves or one another. And when this happens they break at the point of intersection, only to join up again instantly in some fresh configuration. Existing loops, twisting into figure eights, split in two. Infinitely long strings, repeatedly coiling back on themselves, shed loops continuously.

To begin with, these loops are small. Even squirming at near-light speed, a string takes time to curl back on itself and cross over. The closer it is to the grand freezing, the lesser the opportunity for folding back, and the shorter the resulting loops. Small loops, then, are characteristic of the early universe, larger loops of times more remote from the Big Bang.

And yet this does not mean that space gradually becomes more and more congested with string loops of every size. As soon as a new loop is formed it begins to vibrate wildly in the manner of all of its kind. Just as a plucked guitar string generates sound waves, so a vibrating cosmic string emits rhythmic waves of gravitation. Waves which, in turn, steadily waft away the string's energy, causing the loop to shrink – and eventually to disappear. Whatever its initial size, a string loop can survive for only about ten thousand oscillations. And since smaller loops oscillate faster than larger ones, their lifetimes must be correspondingly less.

Time passes. Space grows. And the complex quivering network of cosmic string evolves. Such is the style of its evolution that the string network retains the same overall appearance. Only its scale changes. Smaller loops die out. Larger ones, as they are born, become the dominant type. And so it goes on. Steadily the average size of loops increases. As does their mass. As does the strength of their gravitational attraction.

And so to a crucial question. When did cosmic string start to affect significantly how ordinary matter evolves? When did it start to make matter clumpy? Not right away. Not immediately after the Big Bang – because then the loops of string were still too small, too lightweight, too short-lived to act as important centers of condensation. Besides, in the very beginning the material in the erupting fireball of genesis was much too hot to be shepherded and corralled into many distinct lumps. Only as the string loops grew and matter cooled could the universe become noticeably clumpy.

And now we watch this process with childlike curiosity. In imagination we hold up the young universe in our hands, as if it were a crystal ball. We turn it around and peer inquisitively into its depths as it evolves through its first few thousand years. And we gaze at the quick-spreading mist of glowing matter, at the dark, shivering veins and loops of cosmic string embedded within the mist. Wondering what will happen next. Knowing that somewhere in this precious everything-globe is our own tiny particle.

Gradually, the glowing mist within the ball changes. Countless microscopic points of more intense light begin to appear all throughout it. And we know that at the core of these bright points there must be a loop of cosmic string. A loop roughly 100 light-years long, it transpires, for this was the critical size of string needed to trigger the local infall of matter in the early universe. As soon as loops of that length were formed they began pulling matter from the regions around them. And slowly, over thousands and thousands of subsequent years, each of them drew together a great shining cloud of particles that one day would become ...

But now we seem to be falling into the cosmic crystal ball. Falling in, head over heels, and shrinking. Shrinking. So that, in a few moments, that toy we had previously held in our hands has become, once again, the real universe, 30,000 years after its birth.

***

Once again we see our particle hero. And we realize now, as if in waking from a dream, that we are inside one of those tiny bright lights that condensed from the glowing mist of the crystal ball. Only now it seems no longer tiny. We have the vague impression of a huge structure taking shape around us. Yet we cannot actually see it. From the vantage point of our proton we cannot see anything but a uniform haze of light. It is as if a great city were being constructed, but all of it in the midst of an impenetrable fog.

Nor is the source of that light-fog hard to find. Looking out from our particle, we notice that the billions of photons swarming around are still highly energetic. Each of them, on average, is still well able to disrupt any would-be atom. And because of this the nuclei and electrons in the huge cloud to which our proton belongs remain in their dissociated state.

But free electrons, it turns out, offer a huge target for any approaching particles of light. An electron simply "looks" much bigger to a photon when it is outside an atom than when it is inside. And so the photons habitually bump into the very specks of matter whose liberty they have helped maintain. Each time this happens – each time a photon scatters off an electron – it adds a little to the overall pressure of light that acts on the whole plasma cloud.

Since no photon can travel far without blundering into an electron, this explains why there are no glimpses of the far-off at this stage. Zigzagging its way across space, light arrives at our proton (and at every other point in the cosmos) from all directions with equal, inscrutable brilliance. Frustratingly, for us eager tourists, the universe is totally opaque.

But not only that. Not only is the light-fog spoiling our view. Much more important, it is influencing the very evolution of the universe. The pressure of light, caused by photons continually bouncing off free electrons, is buoying up the clouds of nebulous matter that have gathered, like cotton candy, around cosmic string. Even as the gravitational force of that string, aided by the self-gravity of normal matter, strives to squeeze the clouds tighter and tighter together, so the outward pressure of radiation attempts to blow the clouds apart. The result: A balance point is reached. The nascent clouds remain intact. But they are prevented from collapsing further by the force of light within.

Years pass. The universe grows. And the cosmic temperature falls. 100,000 years AG, 200,000. Still no change. 300,000 years AG.

A third of a million years into its expansion. And the universe has seen its temperature decline to a mere 3,000 degrees. The dance of the photons is now less exuberant. And slowly, surely, it is becoming cool enough for the first atoms to form.

Not far away a proton and an electron combine. From their marriage – a hydrogen atom. But will it survive?

A photon approaches on collision course, strikes the newly-wed electron, which then absorbs the photon's energy. Higher it jumps, away from its proton partner. Yet the electron is not thrown free. The energy it gained was not sufficient to snap the electromagnetic bond it had established with its tiny nucleus. Simply, the electron hops to a higher energy level within the atom before quickly falling back to its normal, "ground" state.

Nearby another hydrogen atom forms. Again, within moments, its electron is torpedoed by a photon. Again the same result. The photon narrowly lacked the energy it needed to sunder the electron-proton pair.

More and more the atomic population swells. Atoms of hydrogen and helium mingle thickly now with the remaining solo electrons and protons. And with the decline in the number of free electrons, there is less opportunity for photons to scatter. So the radiation pressure throughout the cloud is falling rapidly.

Now it is 335,000 years AG and our proton is no longer alone. It has found itself an electron mate and now moves in close partnership with it through the cooling, condensing mist around them. The pressure of light has eased further, leaving gravity, always waiting in the wings, virtually unopposed.

The long era of plasma is nearly over, and with it the time during which radiation and matter were inextricably coupled. A new age is about to dawn – the age of the atom.

***

As it cools the universe changes dramatically in color. In the beginning, during the first few hours, it was the lurid, alien hue of gamma rays, the most energetic, highest-frequency waves of the electromagnetic spectrum. Then, steadily, the peak of its glow drifted down into the X-ray region, and more recently into the ultraviolet. Some tens of thousands of years ago the main light of the universe began to fall upon that narrow window accessible to human eyes, the most numerous photons around clustering at the violet and blue end of the optical spectrum.

Now, with its average temperature sinking just below 3,000 degrees, the cosmos shines everywhere yellow. Glorious, brilliant, golden-yellow. And with matter at last shedding its opacity to radiation, some of this aureate light is crossing vast regions of space to yield the first true cosmic vistas, the first startling glimpses of the faraway.

Between nearby scudding cloudlets of partial plasma, still resistant to the passage of light, we catch breath-taking glimpses of what lies beyond. For a few moments, or perhaps years, we gaze through these clear, temporary portals in our own super cloud to witness events in the greater world outside. Light-years away, in every direction, other bright mists of light are condensing and evolving. And, in general design, we know that they must be much like the structure in which we find ourselves.

Once more we turn inward to contemplate our own minuscule hero. From its birth out of spacetime to primal energy speck, to quark, to proton, to atomic nucleus it has progressed – even as the macrocosm around it has evolved. Nature, it appears, is growing in sophistication at every level, from the subatomic to the cosmic. Growing in sophistication, complexity, and variety. And revealing, moreover, as times goes by, a most potent, inner creative urge. Where will that irrepressible drive to unfold lead next?

Dimly we sense the movements of matter around us. The streaming of gas. The passage of giant, nacreous shapes. We feel the indomitable force of gravity drawing together the hydrogen and helium of this native cloud, for since the counteracting pressure of light has all but vanished, there is everywhere the impression of compaction and collapse.

More rapidly now does time seem to pass. Tens of thousands of years are as moments to us. Our proton, in its atomic spaceship, is caught up in some great, fast-flowing current. Yet it remains uncertain where that current will lead. Only can we be sure that the local density of gas is rising. That the cosmos, once more, is in a fervent creative mood.

And again our thoughts race out to the universe at large. So that it seems, even as we ride with our particle toward its new fate, we stay closely in touch with the changing appearance of the entire cloud about us. A hazy ball it is from the outside. A golden ball, one of billions, its boundary blurred where it meets and merges with the ambient ocean of sparser cosmic matter.

For a while it seems as if we may have circumscribed the material essence of the universe. There is ordinary matter, mostly atomic hydrogen and helium. And there is cosmic string, about which great clouds of normal matter have congregated.

But even as we wait to see what new inventions nature may bring forth, we sense that something is awry. The movement of our proton is wrong. The dynamics of the whole universe is wrong – wrong, at least, if we allow only for the gravitational effects of conventional matter and of string. There must be some other player, some other wielder of gravity whose presence we have formerly overlooked. And though apparently hidden from view, nonetheless we can be sure now that this third component, this "dark matter," exists.

Like normal matter, and like cosmic string, dark matter must have been forged within the earliest moments of the Big Bang, around the time of the grand deunification at 10^{-35} s AG. But forged of what? Not of baryons – not of protons or neutrons or any near relations of these particles. That much is certain.

But then what does dark matter contain? Neutrinos? Perhaps – but only if neutrinos have mass, and that issue is far from being resolved. On the other hand, maybe dark matter is fashioned of some much more unfamiliar stuff. Depending on the exact blueprint the universe followed, species of particle unmentioned in our tale to date may have flashed into existence at time's threshold. Among these exotica, these prime candidates for dark matter: gravitinos, photinos, and axions. Each a new, conjectured type of particle. Each with its own unique attributes, the details of which are not important to us – except in one regard. That is, the nature of these particles is such that they shun interaction with ordinary matter. They cannot be seen, cannot be detected in any way, save by their gravitational influence.

And that influence, we realize now, is very, very powerful. So powerful that to explain the way the visible contents of the universe are behaving it seems there must be at least ten times more dark matter than bright. Which is to say, no less than nine tenths of the cosmos is in a form that renders it entirely unobservable.

***

One hundred million years have elapsed since first we saw our proton. And how much its environment has changed in that time! No longer is the universe simple to describe, its energy and material contents spread more or less evenly about space. Now there are great evolving associations of matter, held in check by their own self-gravity, in which the density is much higher than average. And between these gatherings there are even larger gulfs where the density is low – and becoming lower still.

With vision divided between two levels, we continue to watch nature's progression – from the viewpoint of our hardy proton and from a sweeping global perspective. Before us the microcosm and the macrocosm are simultaneously arrayed. And between each breath a million years go by.

Steadily we begin to discern the import of dark matter. We begin to appreciate that whereas cosmic string may have triggered the development of large-scale structure in the universe, it was due mainly to dark matter that this development could continue.

In hosts far greater even than those of the protons, neutrons, and electrons, the mysterious particles of dark matter assembled around cosmic string. At first these two very different subatomic creations intermingled. But then, gradually, the constituents of bright matter and of dark began to segregate. The particles of ordinary matter started to drift inward, to slide down the steep slopes of a valley – a gravitational valley – sculpted by the encompassing cloud of dark matter. For now gravity is revealed to us in this new and more lucid way. Whereas before it had been an unexplained "force," broadcast somehow by any object with mass, now we can comprehend gravity in terms of the very geometry of space and time.

Spacetime, so our earlier voyage to genesis revealed, is comparable to an elastic skin. Empty, devoid of matter, its strange supple surface is smooth and flat. But place objects upon it and the elastic skin of spacetime stretches. Each material thing, from the humblest subatomic particle to the mightiest loop of cosmic string, creates its own depression in the spacetime skin around it – the greater the mass, the greater the resulting warp. Any other mass traversing the curved sides of such a warp will have its trajectory altered, just as a rolling marble is deflected from its original course if it comes across a dip in the floor. Seen at a superficial (human!) level, the ensuing shift in an object's path would naturally be taken as evidence for an invisible "force" – a force of gravity. But with our awareness now focused on the mutable topography of spacetime, we grasp a deeper truth – that mass determines the local geometry of spacetime. And that the details of this geometry, in turn, influence the motion of other objects nearby.

Massive indeed is the great cloud of dark matter that now surrounds our particle and all of its sibling nuclei of hydrogen and helium. To our sensitized vision, the spacetime valley created by this unseen cloud appears like the inside of a giant bowl, the particles of ordinary matter around us like tiny beads spiraling down the sides of that bowl. Racing to the bottom of the bowl – the floor of the spacetime valley – where they will meet and unite.

Faster and faster comes the succession of events. Our own proton, with its electron companion, is moving ever more swiftly, part of a vast, warming, thickening vortex of matter. And now, within this giant, rotating cloud, this cosmic tornado, many smaller features are appearing. Streaks and swirls, self-gravitating knots and eddies are forming, breaking up, reforming, out of the haze. Held captive by their beauty and prismatic variety, we fail for a moment – for a few million years – to grasp what is happening. Then realization dawns: Deep within these condensing nebulae, the first stars are about to shine. Realization widens: The emergence of stars heralds a new stage in the development of the supercloud around us.

A momentary clearing opens between the gas and dust and glowing stellar delivery room. And through it we look beyond. Hundreds of thousands of light-years beyond. To see, by the uncounted millions, other misty, still-forming cities of stars. All across space the galaxies are being born.

***

All across space?

No, not quite. Nature would never simply daub the galaxies at random throughout the universe. Complexity, subtlety . . . and beauty – these lie at the core of all evolution. "Expect the unexpected!" could be the cosmic motto. The embryonic galaxies, we see, are arranged not haphazardly but in great winding ribbons. Ribbons millions of light-years long. Slender filaments, galaxy-sequined.

And of course it had to be that way. Now it all begins to make sense:

Three things were created just after the Big Bang: ordinary matter, dark matter, and cosmic string. Each, for many years, evolved independently of the others – concerned, as it were, only with its own development. There was little or no coupling. Ordinary matter spread and cooled and transformed by stages into atoms. But left alone it would have thinned and thinned endlessly. Dark matter, whatever it may be, spread and cooled likewise. And by itself, too, would have been hopelessly scattered by the outblast of genesis.

But then there was string. By its very nature, string was immune to dissipation. Ten thousand years after the Big Bang the space-wide network of cosmic string looked in essence the same as it did half a second after it was formed. It had simply grown bigger – in exact proportion to the increased size of the universe. And that property of self-similarity was vital. Because it meant that string could provide a permanent anchor, or framework, for matter to hold on to.

At first the loops of string being generated were too small, too gravitationally feeble, to counteract the explosive power of the Big Bang. But, in time, longer and more massive loops appeared. By the time the universe was several centuries old, loops big enough to start the slow nucleation of the galaxies were in existence. Only about 100 light-years long were these curious galactic seeds. Yet their gravity was compelling enough, intense enough, to draw matter in from their surroundings. As this matter was captured, its own gravitational force fell in league with that of the string. Which was essential. For as the years went by, and the pregalactic cloud of matter gathered, the string seed within it steadily withered. A piece of cosmic string 100 light-years long can last for only about a million years – a trifling instant in the lifetime of a galaxy. But after that first million years enough matter condensed to be able to continue the task of galaxy-building unaided by string.

Strange had been that symbiosis. Matter needed string to rally it together and would have progressed no further without it. String, on the other hand, would have served no purpose without matter. On its own, string can produce nothing but more string. Strange, too, that all hope for a more interesting universe to come should rest with what were, essentially, incidental defects. But then this whole reality is beginning to look more and more as if it is exquisitely intertwined, its various parts entirely dependent upon one another.

Consider, too, the relationship between normal matter and dark. Both these types were lured, indiscriminately, into the pregalactic clumps by cosmic string. Dark matter dominated – perhaps by a factor of ten or more – over its visible counterpart. Dark matter appears incapable of evolving any further by itself. But once the binding power of string had gone, the gravitational influence of dark matter was to prove decisive in ensuring the continued collapse of hydrogen and helium. Without ordinary matter, dark matter would have had no meaning. But in the absence of those vast, enveloping clouds of unseen and unseeable material, all the infinite variety of forms that atoms and molecules and planets and life can take would have been unrealized. Interconnections. Always, interconnections. Nothing in nature, it seems, exists without a purpose.

And those ribbon-like clusters of galaxies before us now? They are readily explained. For of course the string loops never stopped growing in size. We overlooked that while our attention turned inward to the development of our own protogalactic cloud. But now, as we stare through that ragged gap in our new-formed galaxy, we see clearly the filamentary arrangements of the galaxies at large. Why are they so? Because they, like the individual galaxies, have built up around cosmic string. Only in this case the loops involved were bigger. Very much bigger. Perhaps millions of light-years long.

Even as the galaxies were taking shape, neighboring protogalactic clouds fell collectively under the sway of lengthier loops of string. Several clouds clustered together along a filament. And then, as even longer loops evolved, these fledgling groups of galaxies were themselves drawn together to create bigger clusters. Clusters that to us now seem like fabulous shining rivers and tributaries coursing this way and that, spanning the light-years.

The birth of stars. The birth of galaxies. The emergence of great galaxy clusters. All this is taking place simultaneously as we look out on the universe of about 150 million AG. Matter everywhere, on every scale, is developing with vigor.

But too long have we been seduced by such grandeur. Preoccupied with the overall evolution of the cosmos, we have neglected our own trusty particle. How far has it travelled in our absence? How many new suns has it seen beam out their first, triumphant rays? This young galaxy in which it moves has become a true metropolis of stars since last we surveyed it. Billions of stars have already condensed and blazed forth within it. Many more are incubating deep inside the huge dusty cloud banks that throng the galactic interior.

And now into one of these great star-making machines our intrepid particle is tumbling. Not since the earliest days of the universe has it been engulfed by matter this dense and warm. But whereas, in the wake of genesis, particles everywhere were flying frantically apart, now, inside this nebular womb, matter is pulling itself confidently together. Over millions of years denser-than-average clumps assert themselves throughout the cloud. Then, over millions more years, these richer associations self-gravitate further into smaller, denser, warmer globules – star embryos.

Too close to such an object has our particle ventured now. So that it has become prisoner of the globule's gravitational field. Deeper into this congealing dark the particle plunges. More and more thickly other atoms of hydrogen and helium, along with fine grains of dust, swarm and stream around it. And with the rise in density comes a rise in temperature. Until after a few more million years ...

Star birth!

From somewhere below – from the core of the globule – hot blue light is suddenly erupting. Apparently the central yolk of this collapsing stellar egg has become so dense, so hot, that now it can generate its own intense outpouring of energy by the "fusion" of hydrogen into helium. As to the details of that process, they can await description. And in any case our particle is not directly involved with them. Not in the Dantean core, but in the relatively cool, upper atmosphere of this newest of stars our hero finds itself. Quickly stripped of its solitary orbital electron in the first searing blast of stellar radiation, it roams within the topmost few thousand miles of the star as a naked proton once more.

Like a newborn baby, the infant star coughs and sputters, as if struggling to win mastery over its fiery core. Brilliant but erratic torrents of light, visible and ultraviolet, pour from its surface, rush into the surrounding space, and break upon any unconsolidated gas nearby. So is this residue of the parent nebula set aglow by the light of its progeny. Soft red and blue shines the circumstellar cloud – some of its radiation owing to starlight glanced off dust grains, the rest due to the glow of recombining atoms, like those in a neon tube, temporarily parted from their electrons by the star's blazing ultraviolet.

But not only does light from the new sun decorate the surrounding nebula in glorious hues. It also sends powerful shock waves through the nebular gas, waves which, as they travel out, compress the gas and eventually help trigger the formation of other stars.

Perhaps, too, in due course, this star that harbors our proton would have nourished its own family of planets. Already some icy globules are condensing out of the swathes of gas and dust that remain in attendance. But how those worlds might have fared in the fullness of time, how they might have evolved under other circumstances, we shall never know. Because, even as they take shape, they move quickly toward their doom. Soon, terribly soon, a cataclysm will befall this pristine sun and its environs that will tear it apart completely.

***

How could we have known that as our particle journeyed around and around this nascent galaxy, its path had it ever closer to the galactic heart and to the beast that dwells there. How could we have known that only a thin veil of obscuring dust lay between its newfound stellar home and the lair of that impossible monster?

But now the veil is lifting – no, is being savagely wrenched away. And at last the creature at the core of this youthful star city is revealed.

Our shock on seeing it is like that of walkers who, as the mist around them suddenly clears, realize they have been strolling at the very edge of a precipice. Except that from this precipice there is no shrinking back.

Directly ahead, spanning half the sky, is a flat, whirling disk. Slightly edge-on we see it. A hot, fiery disk, composed of what at first looks like molten lava, until the scale of the thing bursts in on our stunned senses. Those lumps in the outer reaches of the disk are not melted rocks. They are whole stars, their contents being pillaged, their planetary systems ripped from their grasp. Farther in, nearer the hub of the disk, the material is even hotter, smoother, spinning more rapidly – here the stellar remains are beyond recognition. And at the disk's very center? There lives the dark ruler of this hellish realm, a source of gravity so tenacious that it will let nothing, not even light, escape from its clutches. There, the star wrecker – a supermassive black hole.

How had it come to be here? This bottomless pit in space-time. This aberration of nature, less than a light-day across yet with the mass, and gravitational pull, of a hundred million suns.

Perhaps it was formed, along with many others of its kind, in the first instant of the Big Bang. Black holes great and small might have been bred in huge numbers at that extraordinary moment, when matter was so dense that even the slightest irregularity ran the risk of vanishing down a gravity chasm of its own making. But then, if this monstrous black hole really is so ancient, what relationship could it have had to the cosmic string that seeded the galaxy? How did the string and hole come to be associated? And why was the string not simply swallowed up by its even more gravitationally severe partner?

More likely, it seems, the core black hole came later, long after the string had wasted away. Perhaps it was born only when the protogalaxy had acquired most of its mass and the bulk of its ordinary matter had begun to condense at the center of the encompassing cloud of dark matter. As millions of stars' worth of hydrogen and helium started to pile up at the galactic core, it could well have reached some critical density at which it condemned itself to ultimate collapse. And so may the dark heart of this fetal galaxy have come into being.

For stars well distanced from the galactic center, the demon that lurks there poses no threat. From afar the black hole's gravitational field – the curvature of its spacetime crater – seems no different from that of any equivalent mass. Just as the Niagara looks like any other big river. Until it hurls itself spectacularly over the Falls.

Too late is the terrible secret of the black hole revealed. Namely that, beyond a certain point, the spacetime walls surrounding the hole plunge sheer to infinity. Once over the brink of those fearful cliffs, there is no hope of escape. No hope for the tortured, blazing matter in this strange whirlpool before us. Already it has been drawn in beyond the point of recall.

And now the young star that shelters our proton is coming to the fringes of the vortex. Is being tugged and stretched, stretched in to an egg shape, by forces it cannot resist. Other stars from the same nebular brood have already succumbed to this gravitational assault. Their gassy contents are smeared in great burning arcs around the rim of the swirling disk. And suddenly our particle, too, is cast into that infernal whirlpool. Its star has finally given up the unequal struggle and been dismembered. Looking out to the edges of the spinning disk, we see other stars and gas clouds preparing to meet the same fate. Preparing to become fodder for the insatiable beast at the core.

How big is it, this whirlpool of plundered star stuff? Ten light-years across? One hundred? Somewhere between the two perhaps. And why is it so hot? The black hole gives off no heat. Gives off nothing at all. So why should its meal burn so brightly?

Those parts of the disk closest to the black hole, we notice, are spinning the fastest. At the periphery the rotation speed is very much less. (Just as the outer planets orbit more slowly around the sun than the inner worlds.) Because of this differential rotation, a tremendous amount of friction occurs between neighboring parts of the whirlpool. The outer layers try to slow down the inner layers, the inner layers to speed up those farther out. And as a result there is a great release of energy as the competing factions slip and slide against one another – energy ultimately supplied by the inexhaustible gravitational reserves of the black hole.

Now that process is graphically revealed to us. In the midst of a small clot of ex-star matter, our particle careers on its terminal flight toward the central hole. More and more furious becomes its chase around the wheeling vortex. Faster it goes, and deeper. Becoming hotter. Faster yet. And as the doomed material around us gathers speed, we see very clearly that some of its frenzied energy of motion is being passed on, as frictional heat, to gas farther out in the disk. So is there a continuous passage of high-grade energy from the fast-moving inner parts of the disk to the more sluggish-moving outer parts. And thus is some of the black hole's gravitational energy converted to heat and light. By proxy the darkling beast shines.

At the outer brim of the accreted disk the temperature had been only a few thousand degrees. Hot enough to make the captured gas there simmer in the infrared and visible regions of the spectrum. But as our particle is swept inward the temperature climbs. To tens of thousands of degrees it climbs. So that matter here glows white and violet and, eventually, ultraviolet.

And now the beleaguered proton has come almost to the threshold of the black hole itself. To the innermost zone of its vassal domain. Where the temperature is ten million degrees, and matter, in its death agony, lets out a final scream of X-rays.

And then?

In a few days it will be the end for our particle. Once it has crossed the invisible barrier known as the "event horizon," which immediately surrounds the black hole, it can never again return to the normal universe. The event horizon is like a one-way valve to . . . other spacetimes? Total obliteration? Whichever it is, no news of what takes place within a black hole can ever be sent back out. The black hole is truly, definitively black.

Our proton begins its last trip around this wild carousel. In just a few hours from now, along with other ill-fated matter around it, it will penetrate the event horizon and . . .

Collision.

A chance bump with a neighboring particle has abruptly altered the proton's trajectory. Now it is hurtling out of the whirlpool – seemingly on a direct collision course for the black hole.

Before us the monster gapes, like the mouth of a colossal, round cave, impenetrably dark. But. Our hero is heading, we realize now, not for the center of that blackness. Instead, it is veering up. Veering sharply up and away from danger. With luck it may skim the event horizon without actually crossing it. Up, up.

To safety.

And at last we take time to marvel at the scene of destruction around. Below: the fervent whirlpool, ablaze with radiation from X-rays to infrared. At the outskirts being joined by sundered stars and tattered clouds of gas and dust; at its inner margin feeding the core monster with looping, licking tendrils of multimillion-degree plasma.

And ahead? Ahead, the escape route.

Two points: This black hole spins (as perhaps all such objects do). And it is not only matter that it lures and bends to its will. Magnetic fields, as well as more tangible stars and interstellar gas, are being sucked in by its powerful gravity. As the black hole scrambles the incoming magnetic field lines and drags them around at high speed, so it operates in the manner of a superscale dynamo. Via the turbulent magnetic field, the black hole transforms some of its spin energy into a vast electrical current. The outlet for that current – two narrow channels running north and south along the hole's spin axis. The escape route.

From each of the poles the black hole sprays a hugely energetic fountain of charged particles. And it is along one of these that our fortunate proton is now moving. Moving at very close to the speed of light. So that within just a few thousand years it has risen far above the plane of the galaxy.

Looking down, we have a fabulous bird's-eye view of that great assemblage of stars we are leaving now forever.

**

Yet the view in retrospect is different – remarkably different – from what we had expected. For we had not reckoned on the enormous contribution of the central supermassive black hole to the galaxy's total energy output. That contribution, coming mainly from the sizzling vortex of matter around the black hole, exceeds by a factor of a thousand the combined radiation of all the stars in the rest of the galaxy! Five hundred billion suns burn within this stellar commune, its luminous bulk sprawled across 150,000 light-years of space. But the misty, mellow glow from that stellar host is almost completely drowned by the searing glare of the core.

Hundreds of millions of years go by. The cosmic clock advances to one billion AG. And as we move ever farther from it, the hazy peripheral structure of the galaxy fades from sight, leaving behind just the intense central point source. More like a single, outrageously bright star it looks now. And so we call it by a new name – "quasi-stellar object." Or simply, "quasar."

How many other young galaxies are also passing through this extraordinary phase? Quite a number it seems. From the scattering of similar brilliant points across the sky, we might guess that at least 1 or 2 percent of all galaxies in the universe harbor active quasars at this stage. Within these brightest of galactic beacons there must be giant black holes much like that from which our proton narrowly escaped. Yet the quasar phenomenon depends not only the presence of such a beast but also on the availability of star food nearby. Where there is plenty of matter in the galactic core to stock its fiery whirlpool, the central black hole can raise the light of hundreds of trillions of stars. But in time that immediate, rich supply of nourishment must diminish. In time the innermost parts of the galaxy must become clearer of gas and dust and stars, so that gradually the black hole will be starved, its quasar light dimmed.

What we have styled "quasar," in fact, surely marks only the upper bound of a complete, unbroken range of galactic activity. Some galaxies, perhaps the majority, may never have shone as true quasars. Perhaps at their cores lie more modest black holes (or none at all). Or perhaps they could never stoke their central engine rooms fast enough. Even so, looking around this still-immature universe, we observe that the inner region of almost every galaxy shines with unexpected brilliance. With too much brilliance, it seems, for normal starlight to be the source.

Again hundreds of millions of years go by, and that suspicion is gradually confirmed. The true quasars, by and large, are growing fainter (though occasionally one will flare up, as if it has found a fresh, temporary wealth of star fuel). And as they fade the quasars begin to appear more and more like conventional galaxies. Their superluminous cores dim, their vastly more extensive stellar suburbs come into view. And, on a lesser scale, the same happens to other galaxies. After an adolescence spent not always in the pursuit of staid convention, they settle down to a more sober adulthood.

We gaze all around at the variety of galaxies. Over millions and millions of years, as our proton coasts on alone through the void, we come to know every species of the galactic zoo. And we learn that not only in the scale of their core activity do star cities differ. There are many diverse shapes and sizes of galaxy. Some are round or ovoid with few internal markings. These "ellipticals" have used up virtually all their supply of loose gas and dust in fabricating stars, so that now they are little more than tidy arrangements of suns and space.

By contrast, a second great class of galaxy, the "spirals," shows dazzling variety of form. The inner bulbous hub of a spiral resembles an elliptical in miniature. Here the stars are evidently very old, so that they must have formed early on in the galaxy's history. But spreading out tens of thousands of light-years from this ancient hump is a much broader and flatter disk, bedecked with interstellar clouds and bright youthful suns. Within the disks of the spirals, winding density waves like ripples in a pond propagate outward. And, as these waves pass through regions rich in gas and dust, they stimulate the interstellar material to collapse into new stars, which then illuminate great curving arms in the disk and give the galaxy its name.

By why, we muse, have some systems become spiral, others elliptical?

Perhaps, as they formed, spirals spun around so fast that they were unable to gather up all of their star-making material into a single ball as the ellipticals did. Instead, part of their contents was whirled out as a broad, gassy pancake in which star building could proceed only at a much slower rate. How fascinating it would be to be able to explore one of these beautiful, complex systems in more intimate detail – from within.

How fortunate that now, it seems, we may have just that opportunity.

For five billion years our proton has been adrift in intergalactic space. And in all that time it has managed to weave its way between the slender islands and atolls of galaxies that populate this universe, avoiding an encounter. But now that lonely, lonely voyage may be coming to an end. Only a million light-years away is a glorious Catherine wheel of light, a spiral galaxy of over 200 billion stars.

Closer and closer it looms, spreading impressively across the sky, its brightest suns and largest nebulae making themselves individually known. Closer and closer. Entering the galaxy's material realm now, our proton sweeps past the central bulge on an arcing path that plunges it into the outskirts of the disk. Ahead, for the first time in five billion years, lies interstellar space.

***

We had forgotten how well separated are the stars in a galaxy in comparison with their size.

Take fifty oranges. Scatter them throughout a sphere 10,000 miles across. That is a measure of the seclusion of neighboring stars in a galaxy. Relative to the size of object, interstellar distances are very much greater than intergalactic, though in absolute terms they are thousands or even millions of times less.

Still remote these tiny, cold points of light around us may be. Yet at least the time scale over which the background appreciably changes is now much reduced. As our proton-explorer travels on, the stars seem to drift steadily past, whereas, over similar periods before, the galactic landscape had remained almost static. Ahead, new stars and nebulae emerge out of the blackness, while others behind fade from sight. Often we imagine that our particle is bound for a close rendezvous with some upcoming star or star cluster. But then we recall the huge distances between everything, and our hopes for a near miss disappear.

The stars themselves are minute targets, almost impossible to hit or graze. But not so the sprawling clouds from which they form. Inevitably, after many millions of years wandering alone down empty corridors of this new galactic home, our particle joins again a great party of gas and dust. A larger than average nebula this, and from its substance larger than average suns are being made. Giant stars are in their embryonic stage here, and when they eventually shine they will be as searchlights amid the candle glow of more common suns.

For billions of years our proton has seen other matter only from afar. For eons it has been merely a wistful spectator to the performances of galaxies and, later, of stars. But now its role is about to change most sensationally.

Within the very heart of a new star's fusion-powered furnace the proton has taken up station. From every direction other matter in the prestellar nebula is flowing in toward it. It is at the center of one of the biggest infant suns in all space, a 30-million-mile-wide blue supergiant.

Crazily it darts to and fro as the fusion fire of this great new star ignites. Violently it smashes against countless other nuclei, densely packed and fast-moving, all around it. And now we see vividly the fusion process in action.

Under the extreme conditions in a star's core, lighter nuclei, as they collide, occasionally stick together to make heavier nuclei. The intricacies of such reactions are many, and they need not detain us. The crucial point is this: Lighter nuclei when lumped together weigh slightly less than when they are apart. That excess mass is liberated, during the fusion process, as a vast outpouring of energy. Only a few million tons of matter "burned" in this way each second suffice to power a modest sun for billions of years.

To begin with, and subsequently for most of its life, a star derives its energy by turning hydrogen fuel into helium. No other fusion reaction involving heavier materials comes close to being so rich a source of heat and light. So, we might expect, big stars, having more hydrogen fuel, ought to be able to shine for much longer than smaller stars.

But no! In fact, that is not the case. On the contrary, a supergiant star squanders its core hydrogen reserves extravagantly. In only a few million years the hydrogen at its center is gone. And then the star must resort to burning helium. And then carbon. And then still heavier substances. All the while its core grows hotter and denser, because only under progressively more and more extreme conditions will the heavier elements catch fire.

We see it happening. Five million years after this great star began shining, our proton is no longer a solitary hydrogen nucleus. By incremental stages it has become fused into a silicon nucleus along with twenty-seven other protons and neutrons. The star's core is now almost full of silicon. And already this almost worthless fusion fuel is being burned, under new, enormously high temperatures and densities within the core, to make iron. To the giant star the start of that process is like a death knell. Iron is the absolute endpoint for a massive star. To fuse iron actually takes more energy than the fusion reaction gives back. So when the core of the supergiant becomes clogged with the nuclei of iron, it is doomed.

A few thousand years later the moment approaches ...

Within less than a second the core shuts down, collapses under its own weight, bounces back because it can collapse no farther, and explodes outward. Then – SUPERNOVA!

In an explosion whose brilliance transiently rivals that of the entire galaxy, the old star blasts itself apart. And now, finally, born in that instant of stellar eruption are substances heavier even than iron. At last the energy is unleashed to create cobalt, nickel, xenon, strontium, platinum, and uranium. Dozens of other elements, too, that in the normal course of stellar evolution can never be made. Only now, in the unshackled fury of a supernova explosion, are they wrought.

And yet most of what spews out from the shattered star are the lighter elements: hydrogen, from the unburned outer layers; helium, from deeper in the old sun, and, in much less amounts, nuclei from carbon to iron.

Why then do we even trouble to mention the creation of heavier elements if they are formed only in minute traces? For this very personal reason: Our own proton has become incorporated into one of these exotic nuclei. As it fled the inner core at one-tenth light-speed, as part of an iron nucleus, it was overtaken by an even faster-moving fireball of energy. In a split second, as the blast wave surged past, it was fused and fused again with other nuclei in an orgy of element-building. Out of the intensely destructive-creative moment it emerged, changed once again – framed now within an atom of gold.

Chapter 4: Morning Star

This world was once a fluid haze of light,
Till toward the center set the starry tides,
And eddied into Suns, that wheeling cast
The Planets.
— Alfred, Lord Tennyson

Two billion years have passed since the massive star erupted in which it was forged. And in that time the gold atom, sanctum to our proton, has travelled far. Eight times around the galaxy it has trekked, weaving in and out of the realms of contrasting suns, bobbing and drifting at the whim of galactic currents. Straying into a quieter backwater of a spiral arm. Only to be swept up once more and borne along swiftly on the crest of some interstellar tsunami.

Yet, even as we watch, these wanderings are coming to an end. The gold atom, with its special cargo, has been washed up on the shores of a still-forming island – a new congealing cloud of gas and dust. And now, in this one virgin outcrop of the universe, nature is poised to take another remarkable evolutionary stride. To begin, in fact, the most astonishing phase of its development so far.

***

Steadily the interstellar cloud grows as more and more itinerant atoms come under its gravitational sway. So does its power of attraction increase, enabling it to add still further to its stock of captured material.

Over millions of years this impromptu assemblage evolves, its tenuous body building up out of particles garnered from all corners of the galaxy. Our gold atom passes just yards from an atom of carbon that, ages ago, dwelt in the methane sea of a world now long-destroyed. Moments later it nearly collides with a hydrogen atom, unchanged from the dawn of time.

And so, in the deep interior of this burgeoning, fomenting haze, knots and eddies, begin to form. Some quickly dissipate, like smoke on the breeze, lacking the mass to hold their contents together. Others are more substantial, so that they succeed in further compacting themselves through their own gravitational embrace. Into one of these denser nucleations our atom of gold has fallen, and now it is slipping, sliding, beyond all hope of escape, into some deepening gravity crater within.

At the heart of that crater lies a star seed. Small and globular it is, its core density and temperature climbing fast. And now our atom is plunging in toward the seed, part of a great river of gas that is streaming in like an excited crowd hurrying to catch a glimpse of some wonderful imminent event.

Suddenly, or so it seems in contrast with the preceding eons, the thing happens. From the very center of the infall, light and heat begin to pour out. A star is born. And now, as if its enthusiasm had all at once given way to panic, much of the inward-rushing gas turns about and starts to flow rapidly away from the central brilliance. In truth it is the keen wind of radiance from the infant star that drives off much of the surrounding nebula. Drives it off, and simultaneously illuminates it in a glorious bouquet of color.

Yet not all of the circumstellar cloud is so dispersed. A fraction of it remains, to settle as a broad, rotating disk around the new star. And, as fate would have it, it is within the inner portion of this spinning, gassy ring that our gold atom now finds itself.

Even as the newborn star splutters and flashes, its nuclear fire still burning inconsistently, the pancake of loose debris around it begins its own process of consolidation. In makeup the disk is like the star – composed mainly of hydrogen and helium, with only the barest sprinkling of heavier elements. Now these various types of atom start to come together on the surface of tiny grains, which in turn aggregate to form larger objects.

In those parts of the disk well removed from the fierce glow of the young star it remains desperately cold. Cold enough for substances such as water and methane and ammonia to freeze out as tiny crystals of ice – interstellar snowflakes in a whirling blizzard. Gradually these snowflakes merge, along with specks of rock and metal, to make icy planetisimals, which in turn accrete and grow. The largest of the bodies thus formed acquire enough gravitational strength to begin sweeping up huge amounts of hydrogen and helium as they orbit, so swelling their bulk and mass many times over. Four such giant worlds take shape around the young star, the largest spanning almost 90,000 miles.

Very different, though, are conditions within the inner portion of the nebula. For here it is much warmer. So warm that the lightest substances must remain as gases, unable to freeze out. The only solid grains that can survive and meld are rocky and metallic. And as these heavier, less volatile flecks come together, they naturally give rise to smaller, denser planets. Planets, moreover, that are destined to stay small, for the temperature of the nebula around them is such that they can never aspire to catch and retain great atmospheres of hydrogen and helium.

Ninety-three million miles out from the youthful star, the largest of these dense worlds assembles itself. As it nears the end of its growth the weightiest of its captured matter – elemental iron and nickel – begins to sink down to form a superthick metallic core. Around this, various lighter rocks accrue, rich especially in oxygen and silicon and aluminium. But heavier elements, too, are here. Elements including gold, one atom of which bears our special proton. For this now is the new home of our particle – a small, embryonic world, third out from an average-sized yellow star. A world that, by some of its own future inhabitants, will become known as the Earth.

***

It is dark when our gold atom makes planetfall. For millions of years it has been dark here, all light from the sun obscured by a choking monsoon of dust and rock that pours down relentlessly from the sky. No succession of night or day is there. No seasons. Only this black, swirling, raging storm of fine, warm particles, in the midst of which the Earth is taking shape.

A million years pass, and another, and yet another. And now, very slowly it seems, the dust storm is beginning to subside. A deep rubicund glow, changing by degrees to brighter orange, starts to permeate the dispersing nebular cloud. Dimly at first, but then more clearly, the globe of the sun shines through, followed later by some of the more brilliant stars around. Superbly luminous are these celestial points of light – white and yellow diamonds in a clearing ebony sky. They are the sister stars of Sol, hatchlings from the same stellar spawn. Their common origin has ensured their proximity in each other's skies – but for only a brief while. Soon this close family group of a few score modest stars is bound to fragment. There is not the gravity within it to

61

prevent its individual members from breaking free and joining the greater community of stars in the galaxy.

Nearer to hand, other changes are taking place. The nascent sun is jettisoning gas from its surface at a prodigious rate, blowing a stellar hurricane that scours the surrounding space of any remaining atomic debris. Half the sun's original mass is shed over the next few million years, returned promptly to the interstellar void whence it came.

Meanwhile, the Earth is slowly emerging. Bright as a coal fire it gleams, its rocks made molten by energy sources both within and without. From inside comes the steady heat of radioactive decay, as elements such as uranium and thorium and radium split apart into stable forms like lead. From outside comes the continuing heat of bombardment.

For several hundred million year after its formation, the Earth is showered with all manner of rocky detritus that still litters the infant solar system. Most of the impactors are trivial, mere dust motes or pebbles or boulders. But a few are much larger, as broad as mountains. And when one of these giant meteorites strikes, the result is sensational. From around the collision zone a towering, circular wall of magma rises up, twenty miles or more, before collapsing to the surface again in a great, blazing splash. Outward from the impactor point, ripples spread sluggishly through the surrounding rocks, dying away only after thousands of miles. And then, perhaps a century later, the spectacle is repeated.

Gradually, the surface layer of rocks cools. Glows more dully. And, in places, congeals. So that, as time goes by, the Earth's magma oceans become increasingly covered by rafts of solidifying matter. Rafts that, over tens of millions of years, join together to form a scum-like crust, steam-shrouded, broken only here and there by shrinking pools of red-hot magma. Overhead, through the mist of rising steam, a black sky, a glaring, yellow-white sun, stars, and ...

A world – half the size of Earth! – closing in. Fast. Growing. Filling the sky, every detail on its alien surface becoming horrendously visible in that last split second before ...

Billions and billions of tons of magma are soaring spaceward in an orange fountain 2,000 miles wide. The Earth has been hit by this renegade world. A savage, glancing blow. A blow that it will survive. But no longer will it move alone in its orbit around the sun. Even as the intruding world vanishes again into the blackness of space, the formidable mass of hot rock gouged out of Earth's mantle is spreading and settling into a dense, dark ring some 400,000 miles across. Over the next few tens of millions of years the fragments in that ring will reassemble themselves. A lost portion of the Earth will reform as a new, independent body – the moon.

***

Now it is 600 million years since the Earth condensed. Ten billion years since the Big Bang. The torrential rain of space debris has eased to a drizzle. And the third planet of Sol is preparing to evolve further in a most extraordinary way.

Driven out of the rock in the mantle and crust by the extreme heat are huge quantities of water vapor, and of carbon dioxide, methane, ammonia, and hydrogen. Through gaping vents and cracks and volcanoes peppering the surface, these gases escape and begin to collect around the Earth as a dense, noxious atmosphere.

Over millions of years the fugitive water vapor gathers as a layer of gray clouds. Becoming thicker. Becoming a single, monstrous storm cloud that completely envelopes the world.

And now drops are beginning to fall from that dark, heavy-laden sky. Great, fat, raindrops, splashing and hissing on the scorching, hard rocks of the surface. Falling as a gentle shower. As a heavy rain. As a deluge, centuries long. Gathering in warm puddles, in pools, in widening lakes. In shallow seas. And finally in a planet-wide ocean, interrupted only here and there by the peaks of smoldering volcanoes jutting up as steep-sided islands.

Relieved of some of its burden, the cloud layer is becoming less impenetrable. Shafts of sunlight occasionally break through it, outlining the high cumuli in gold, glistening off the new-formed ocean below. Making its wave crests sparkle, its breakers gleam as they crash against the shores of scattered, steaming conical isles and lagoons.

But the sunlight that dances on this primal ocean is as yet unfiltered and raw. Powerful, searing rays of ultraviolet, along with those of visible light, spear through the early atmosphere, splitting apart some of the molecules of ammonia and methane and other gases. And so allowing these to join up in new ways, to yield more complex substances, which then congregate increasingly in the upper layers of the ocean.

Once pure and sterile, the Earth-sea is becoming a fermenting brew of many different substances. An organic bouillon. Infused with energy by solar ultraviolet and by occasional bolts of lightning, this oceanic soup of chemicals evolves. More and more complicated molecules build up within it, by chance encounter, by chance reaction. Amino acids. Chains of amino acids. Proteins. Nucleic acids. Some of the basic ingredients for these perhaps arriving on the surfaces of meteorites and comets, having been fabricated far out in space. Maybe, too, the melding of simple molecules into more elaborate ones takes place, at least partially, on land – on lava-strewn shores or in beds of warm clay. By whatever means, the complexity of substances on Earth increases.

And then, at some stage, a very special kind of molecule appears. It may be an unusual protein or nucleic acid, helical in design. Variations on it may have sprung up spontaneously at many different places, not all at once, some more durable than others. That is detail. What is crucial is the fact that this new breed of molecule, by trial and error, has discovered how to make copies of itself. In that respect it is unprecedented. In that respect it displays the fundamental talent of all life.

But self-reproduction? Surely that trick had been learned by the stars long before. A massive star is born from an interstellar cloud, lives out its life, and then, in a final, eruptive paroxysm, reseeds the galaxy for a new generation of stars. Unwittingly, stars make copies of themselves too.

But they are not perfect copies. The process of stellar replication is a chancy, ill-regulated affair. Yes, in some broad, poetic sense the stars do seem to live and die and recycle the stuff of which both they and their progeny are made. But the same could be said of individual subatomic particles. When two high-speed protons collide they may spawn replicas of themselves (along with new particles of other species). Life, in this sense, is a property shared by all matter in the universe.

And that is a possibility, a perspective, worth remembering. Matter everywhere is alive. In that it can make copies of itself. Moreover, in that it can evolve, through natural selection, into more complex, capable forms. The universe is imbued, and imbues itself, with creativity. What quality more than self-creativity proclaims the existence of life?

Even so the biochemical synthesis taking place on the primitive Earth is unique, is very special. Because never before has there been complexity of this order. Never before, complexity so adept at procreation. Out of the terrestrial cradle something is emerging that is profound beyond all expectation. Here, now, on this paltry third world of a modest yellow dwarf star, nature has begun its most far-reaching experiment.

***

Was there some magical moment when the first living thing appeared? A dramatic instant when a flash of lightning, say, struck the primal ocean and caused two innocuous-looking molecules, by chance to fuse and come alive? Of course not. That never – never could have – happened. For the simple reason that there is no sharp distinction between living matter and nonliving. No molecular Adam. Just as there is no clear divide between loud and soft, between hot and cold, between simple and complex. The juncture at which life begins is defined by us, not by nature. "Life" is a convenient, subjective label, not a measurable property of the universe. How far back do we wish to pin that label? On the amino acids? On strings of amino acids – the polypeptides? On the first self-replicating molecules? Or do we reserve it for still more intricate arrangements of matter to come? The choice is ours. Ours alone. And, for that very reason, we should not take it too seriously.

What is important is that an extraordinary process is going on here on the skin of this half-billion-year-old planet. Suddenly a focusing of ingredients and conditions has taken place that is allowing nature to unfold at a runaway pace. Not since the first second of the universe, and perhaps not even then, has there been change and development and genesis so rapid, so rampant. And yet, all along it seems, matter and energy had this astounding potential. The power of the universe to be creative is inherent, the self-organizing property of matter innate. Only must it await the right time and conditions to manifest itself. Like a seed that lies dormant until, at last, nurtured by soil and water, it grows and flourishes. And, in due season, gives rise to new seeds ...

Everywhere on the young Earth the process is happening. In the surface waters of the planet-girdling ocean. In the brown ooze of the seabed. On the wave-lapped shores of the volcanic isles. new microscopic structures are emerging – tiny blobs and flecks of organic jelly, harboring molecules that can actively manipulate and control other substances in the environment. Through the walls of these biodroplets simpler materials are taken to provide energy, to furnish the building blocks for growth and self-replication. Those structures that are able, swiftly and faithfully, to make copies of themselves tend to survive and spread. Less effective designs, the vast majority, die out or become fodder for their more successful contemporaries.

So, over time, do certain specific kinds of organism rise to dominance on the young Earth. They dine on the rich molecular broth around them. They multiply. They compete with one another. And the best equipped of them emerge, triumphant again, to proliferate further.

Yet, of course, it cannot go on forever. More and more the stock of ready-made organic food is depleted. These earliest of living things, bacteria-like in appearance, are crude opportunists. They raid their surroundings indiscriminately, giving nothing back. What happens once their inherited source of nourishment is gone?

Fortunately, another factor is at work – "random mutation." By – accident? – some of the existing organisms are becoming subtly altered. The cause of their change may be the decay of a radioactive atom. Or an ultraviolet photon from the sun. Or a charged particle from space. It may be almost anything. But the effect is to produce, among the billions and billions of individuals of a species, a mutant, a freak, a deviation from the norm.

Most such changelings prove disastrously inferior, evolutionary cul-de-sacs. Even if they survive themselves, and even if they can reproduce, they yield offspring that cannot compete as effectively as their kin. And so, quickly, the mutation dies out.

And yet. Once in a while a mutant emerges that is actually better in some way than the original. Maybe it can tap a new source of food. Maybe it is more resilient to attack from rivals. Whatever its advantage, because of that slight benefit it is more likely to live long enough to spawn others with the same characteristics. These offspring, in turn, are more likely than average to survive and breed. And so on. Thus is the species as a whole steered in a new, unexpected direction. Thus do new life-forms evolve; by natural selection.

Nor is this process confined merely to biological systems. What is a planet, a star, a galaxy, the universe itself if not a product of natural selection? Nature sows its wide vessel with the germ of every conceivable creation, however unlikely, however outré. Sows it and then leaves these potentialities to interact, to contend and combine, to self-select – to grow, if they can, into adult forms. So do would-be stars compete with one another for gas and dust, the most successful going on to consolidate and mature. By their particular makeup and style of development, these survivor-suns then influence what the next generation of stars will be like. And the same is true of molecules, or planets and continents, of all physical systems, from the subatomic to the cosmic. Different manifestations of matter and energy vie for available resources, including themselves, and the fittest survive.

***

Now the Earth is just over a billion years old, and most of the primordial organic soup in the ocean has been consumed. The bioscape is changing, irrevocably. Becoming more discerning, more sophisticated. And steadily it is beginning to favor those organisms that are not so dependent on having a ready-made diet. Selection pressure is spurring the growth of new creatures that show innovation. That can manufacture their own food. Creatures like the blue-greens – sea-dwelling microbes that have learned how to trap and harness directly the energy of the sun.

Around shorelines all over the globe the blue-greens are swarming. Absorbing sunlight by means of a special chemical – chlorophyll – in their bodies, they furnish themselves with the energy they need to build complex molecules from simple substances like hydrogen in the water. They are the first photosynthesizers. And now for hundreds of millions of years their fabulous new light-catching skill ensures that they will go unchallenged as the most advanced life-forms on the planet.

But not only does the empire of the blue-greens rise and spread to embrace all the Earth. Through its activities it also alters profoundly the environment around it. Most visibly the lime that the colonies of blue-greens secrete builds up and up as rounded stony cushions and columns in the warm ocean shallow. Curious and impressive these "stromatolites," the first large, life-created structures on Earth, may be. But even more significant are the tiny bubbles that form on their submerged surfaces. For these small, silvery beads contain oxygen, released as a by-product of the blue-greens' photosynthesis. And now, as we watch, the bubbles are detaching themselves and bursting, setting free their precious contents. Setting free a gas that, over millions and millions of years, will revolutionize surface conditions on the third planet.

***

Interconnections. Always, interconnections.

We gaze, entranced, at the endless succession of bubbles rising up through the sun-dappled waters. Bursting. Charging the primordial atmosphere. more and more, with this marvelous new gas. And our thoughts begin, once again, to drift back through time. Far back, to the creation of everything. Of matter and energy.

And of cosmic string.

Oxygen ... cosmic string ...

Why do we sense there is some link here, some communal thread spanning billions and billions of years?

Has it to do, simply, with serendipity? With the universe stumbling blindly, mindlessly, upon new structures and materials? Non-designs that, once found, then grow and grow in significance? Until, eventually, they become central to all of nature's future evolution?

Perhaps. Both string and free oxygen do seem to have been created almost fortuitously, almost as an afterthought, in the wake of some other process. String, we saw, arose in the form of defects, imperfections, in that weird freezing an instant after the Big Bang. And yet, although it began as if it were a minor flaw in nature's scheme, string then went on to play a crucial role in the development of galaxies and stars. And of all higher organization. Without cosmic string there would have been no sun, no Earth. No life.

Was it then just a happy accident? Or was the string's emergence, all along, inevitable, essential – part of some grand, convoluted plan?

And now oxygen. Apparently, again, to begin with, a mere waste product – a cast-off of photosynthesis. But a substance, we realize now, destined to play no less a part than cosmic string did in fostering greater order and complexity. Why? Because as the oxygen builds up in the early atmosphere, some of it, at high altitudes, is converted to ozone by the sun's ultraviolet rays. That ozone, even now, is beginning to act as a radiation shield, absorbing further ultraviolet and preventing the most harmful rays from reaching the ground. Meanwhile, mixed with the ocean waters and blended into the atmosphere above, normal oxygen is starting to provide a rich, new source of energy for terrestrial life to come.

Of course. Of course ...

For an instant we see not detail. Not the parochialities of some small, evolving world. We see, we feel, wholeness. Entirety. The universe is a unity. Every star and every planet, every galaxy, every single subatomic particle is somehow interlinked, its identity merged and blended and absorbed into the totality of existence. Throughout all space and time, no event stands alone.

But there is more to it than that. For why is this macrocosm so fertile, so astonishingly creative? Did it have to be that way? Final understanding, we realize, is still beyond our grasp, our voyage through Deep Time still far from complete.

And our particle? Our micro hero? For a moment we catch sight of it, lying there, in burning darkness, under four miles of crustal rock. Waiting. Waiting. We must learn its ultimate fate. We must follow through this stupendous cosmic saga from beginning to end.

And now our awareness is drawing in again, returning to focus on the emergent Earth. To fix once more upon those tiny bubbles that are preparing the way for ...

The bubbles are gone.

The ocean is gone. And through a gap in the lush greenery that now surrounds us, we peer in amazement at a tall, latticework tower. Supported by that tower is a gleaming bullet, two hundred feet high, aimed at the sky, a white mist seeping from its side. On the ground, figures are moving. Small, alien shapes. Life-forms, certainly – but unlike any we have encountered before.

Our cosmic reveries has lasted longer than we guessed. Not for a few moments, but for over three eons we have been gone. Now the Earth is just over four and half billion years old.

And somehow we must try to reconstruct what has happened, what led up to this.

Imagine then:

The time, about 2 billion AD The very photosynthetic power that had brought such success to the blue-greens had now, at last, set the scene for more sophisticated life to appear. Oxygen, escaping from the stromatolites over hundreds of millions of years, had become a major component of the atmosphere. A high ozone layer now served as a barrier against the harshest of solar ultraviolet, making it safe for organisms to dwell at the top of the ocean – and eventually even upon dry land. At the same time rising levels of normal oxygen in the air offered a new source of energy to any life-form that could use it.

The whole ecological climate was changing dramatically. But it was changing now not because of inorganic persuasion but through the action of existing life. The blue-greens were transfiguring the environment that had bred and nurtured them. So that, even as they themselves developed, these tiny primitive creatures caused their surroundings to develop in kind. Where now did life begin and end? At the boundary of the stromatolites? Or was the entire terrestrial biosphere coming alive – emerging as a single superorganism, its many parts implicitly dependent on one another?

***

Too valuable a resource was oxygen to go untapped for long. Over the next half a billion years or so new creatures sprang up that could absorb the oxygen directly and use it to break down organic matter for food. Indeed, so potent a life fuel did the oxygen prove to be that these new "aerobic" organisms could metabolize much faster and so be much more active than the sun-basking blue-greens. Thus began a great parting of the ways. From the aerobes stemmed all of the oxygen breathers, the rapid movers, the food seekers – the true animals. While along a quite separate line, from the photosynthetic blue-greens, arose the more passive, sessile kingdom of plants.

So, steadily, did the diversity of creatures grow. And so did the interplay between different species (and individuals) become more and more labyrinthine. Environments and life-forms developed together and nurtured each other and entwined around each other to yield a dynamic global biosystem of awesome complexity. Cause and effect, life and nonlife became hopelessly intermingled in a Gordian network of feedback loops and relationships.

Nor was it always a case of species in competition. If there were advantages to be gained from it, previously separate organisms would pool their resources and come together – so intimately, so altruistically that in time they evolved into a new, higher form of life. Just as electrons and protons and neutrons had collaborated to form atoms. And atoms to make simple molecules. And so on up the scale. Now this creative hierarchy was being extended into the realm of living things.

Interconnectivity. Self-creativity. The multiform unfolding of matter. These same beguiling qualities of the universe that had been evident since the Big Bang were being manifested here once again. Only now it seemed that nature had risen to new, impossible heights of inventiveness. Like some cosmic Leonardo unleashed for the first time with canvas, brush, and palette, the universe set about painting what for eons it had held only in its fertile imagination. Suddenly its output, its prolific genius, knew no bounds. Two and a half billion years after it had formed, the Earth was strewn with the living sketches of countless different microorganisms. Some lying crumpled and discarded, as if the artist had quickly lost patience with them. Others, those that had better pleased their creator's eye, earning prominence in the terrestrial studio. Enduring. Prospering. And eventually, perhaps, winning a place within more elaborate compositions.

So as the artist learned from his mistakes and formative studies did his mastery grow. Gathering up groups of different, previously autonomous microbes and enclosing them within a common membrane and placing them under control of a central device, nature conceived the first eukaryotes – the first single-celled organisms with a nucleus. Now the mechanism was in place to store and recall a much wider genetic diversity. Within the nucleus could be held the blueprints for trillions and trillions of genetically distinct individuals. And by a new, elaborate process of division, the nucleus could pass on these multifarious plans to subsequent generations. Thus was the way cleared for sexual reproduction to come, and for the even greater proliferation of life's variety this would bring.

It was a little under 1.5 billion BC And now, from this point, the cosmic artist's imagination seemed to run on, untrammeled, almost uncontrollably.

***

Just as groups of microbes had gradually come together and taken on specialized roles within the eukaryotes, so, in time, did the eukaryotes themselves unite. Perhaps through some molecular accident – a chance mutation – the cell membrane of one or two types of eukaryotes became slightly "sticky." Neighboring cells and daughter cells, thus modified, began to adhere to one another. And so did the first colonial life-forms come about. Life-forms like Volvox – a fragile, beautiful, hollow ball, pinhead-sized, built up of many cells, working as a team, beating their flagella in unison, like oarsmen, propelling their corporate self along.

Shades of an earlier time. When naked quarks had been abroad. And we had wondered then how each of these trivial particles could "know" what it was, how it should interact with others of its kind, to form more complex things. Yet still the mystery: What drove these primitive, lone eukaryotes to combine and so harmoniously work together in colonies? Survival? Because a bigger organism was harder to attack and ingest than a smaller one? Because even if several cells in a colony were damaged or destroyed, the collective form might endure?

Inevitably, that was part of it. Natural selection was at work here – as it always had been, as it always would be. But behind the emergence of the communal eukaryotes, just as behind the emergence of nucleons and atoms and molecules, and of all higher forms of organization, there was also the omnipresent urge of matter to assemble and integrate. The individual – the particle, the cell – tended to become part of a greater creation. The quark became part of the proton. The bacterium and blue-green part of the eukaryote. The eukaryote part of a colony ...

And so to 1 billion BC, to the next stage of the hierarchy. For now selection pressure was encouraging the growth of small, accidental differences between cells in a colony – in their ability to extract food or supply movement or sense their surroundings. Nature was promoting cell specialization. And more – the clustering of similar specialized cells into primitive organs. Guts and tentacles and eyes.

Creatures were beginning to appear in the ocean composed not just of cooperating cells but of cooperating, diversified cells. First, the sponges. And then, in ever shortening intervals, more and more complex multicellular forms: jellyfish and sea anemones; polyps and armored trilobites and squid-like belemnites, and, around 550 million BC – fish.

Quarks in nucleons in atoms in molecules in macromolecules in microbes in cells in tissues in organs. In fish!

But was the fish now the individual? Or was the quark? Or was perhaps the Earth itself – the whole planetary biocosm – the only true, indivisible thing? But then the Earth was an integral, inseparable part of the solar system. Which was part of the galaxy. Which was part of the universe ...

The universe. Ultimately, that is what was coming alive. Here in the ocean of Earth.

And not only in the ocean. For in the four billion years since the Earth had taken shape, the volcanic islands had grown and spread and joined together to form a single giant land mass that floated upon the plastic mantle below. Now that ancient supercontinent had begun to break up into smaller continents and islands.

Four hundred million BC And, once again, change. Some intrepid species of fish had begun to alter their habits. To swim into estuaries, and thence farther upstream, in quest of new pastures. Into freshwater rivers and lakes. And not only that. But in the dry season, when their pools evaporated, some fish had learned how to survive on land. How to breathe. How to live not as fish but as amphibians. With lungs as well as, or in place of, gills. And with legs in place of fins.

Quickly now some of these amphibians completed the process, becoming terrestrial reptiles, feeding on the plants that had invaded the land before them. Or, in time, evolving to become meat eaters that preyed upon their grazing cousins or snapped at the giant dragonflies of the Carboniferous forests.

***

Three hundred million BC to 250 million BC And now, steadily, irresistibly, did the reptiles become despotic lords of all the Earth. In size and strength they grew. So that, as dinosaurs, they came to dominate the land, both as plant eaters and as terrifying carnivores. As submarine plesiosaurs and ichthyosaurs they patrolled the oceans. And as pterosaurs, some as big as fighter planes, they suffered only the flying insects as aerial contemporaries.

And yet, even as the dinosaurs slept through the cool primeval night, another, very different type of creature moved unseen around them. Insignificant it was, a shrew-like thing that darted nervously about, taking what it could, grubbing in piles of rotting vegetation in the hope of exposing some larval treat. Its black eyes sparkled, its pointed, glistening nose habitually twitched, and its long skinny tail flicked from side to side. Any sudden noise made it freeze, heart pounding, its whole body momentarily tensed with alarm – until the next instant, when it resumed its frantic search.

As daybreak approached, as the first gray, misty light of dawn seeped into the forest, so this timid animal withdrew to its burrow. It dared not stay out beyond early morning. For then the warming rays of the sun would gradually wake the monsters that slept nearby, freeing them from their bonds of torpidity. It was the one advantage the little creature held over the awesome giants that ruled this harsh, primordial world. It could generate its own body heat and so remain active while the great lizards were compelled to rest. Millions of years before, its ancestor had itself been a reptile. But the competition for food during the day had grown too fierce. There was much to be gained by adapting to a nocturnal lifestyle. So, gradually, through nature's sieve, this one reptile, together with a few others out of many contenders, won control over its body metabolism. It became a mammal.

Yet, warm-bloodedness exacted its own price. To keep a constant body temperature, often much higher than that of the surroundings, called for a good, steady intake of food – animal protein in the form of insects, for example. At the same time, the first mammals fed in the dark, and their prey was small and skittish, hard to find. And they, themselves, were vulnerable. They would survive only if they had keen senses, not only of sight, but also of hearing and smell.

That was crucial, an evolutionary milestone. For the main adaptation to nocturnal living – the dramatic enrichment of auditory and olfactory senses in one small group of embattled creatures – triggered the explosion of the mammalian brain.

In the case of the reptiles, vision was the prime portal on the world. The reptilian eye was a masterpiece of bioengineering. Not only did it detect light superbly well but it helped to analyze the incoming visual data. That analysis, that on-the-fly processing, was done by neural elements within the retina. Sensory cells, in harness with retinal neurons, sifted, refined, and packaged key optical messages to such an extent that very little higher-order processing was left to the animal's brain. The reptile's response to its environment was slavish, its behavior tightly bound to specific stimuli, almost robotic. A thing happened in the creature's visual field; its retinal circuitry raced through some fixed imaging sequence, out of which flashed a signal to the brain; and the animal reacted to that signal, rigidly, predictably.

And, nature being conservative, the first mammals inherited this reptile-like vision. But the dominant night-time senses they then went on to acquire, through natural selection, were hearing and smell. Fresh, extensive neural circuitry was needed to subtend these senses. And yet – this was decisive – there was no practical place for it to go, save one. The inner and middle ear were much too cramped to house several million extra cells for dealing with audition. And so these cells congregated instead in the only place they could, as a new tract of the creature's forebrain. Other cerebral outcrops formed to support the enhanced olfactory sense. So, quite simply, the initial growth of the mammalian brain was in response to a packaging problem.

But the dam had been breached. Torrents of raw data from the outside world came rushing into higher centers of the brain for the first time. Now, through nature's winnowing, those individuals were favored whose brains could most effectively use this data for survival. So were the plans for building better brains cultivated by the species. A "better" brain was one that could synthesize a more potent world model, a higher grade of reality. For the early nocturnal mammals, this meant fashioning an accurate tridimensional sound and scent picture of their surroundings. Only then could the prey be discerned in the gloom, the dark predator uncloaked.

And more quickly followed. With hearing and smell both centered in the brain, it was feasible now to unite the senses. Feasible and enormously desirable. If the sound world and the smell world could in some way be overlapped, merged with the world of sight, then how much richer the creature's reality would become. How much sharper the view of food and foe. And so the brain, encouraged by selection pressure, acquired more circuitry and learned a new ploy – to swiftly associate and integrate the tales told by different senses.

With the emergence of association areas in the forebrain of early mammals, the "realities" of different senses coalesced. Then it was that the brain began to build a more complex world model, a common code it adopted for labeling data gathered by the various senses. Where the labels matched, this served to create something entirely new on Earth – the vague but emergent notion of an object, an object isolated in time and space. Objects, space, time. These were the basic constructs of conscious experience. Here was the wellspring of the mind.

***

Like all else, the senses, the nervous system, the brain were the fruits of nature's random, statistical trials. They had evolved to serve one purpose – to make available those aspects of the environment most relevant to an animal's survival. They shielded the creature from the inundation of all possible knowledge, allowing in only a small selection of information most likely to be useful.

With the reptiles it was the eye that acted as prime censor. No matter how rampant the visual scene around it, the creature's brain received only the most rudimentary stimuli, chosen and packaged by the retinal circuitry.

With the early mammals, though, because of their nocturnal adaptation, far more data from the senses began reaching the brain. Now it, too, could join in the censorship, selecting what was most useful and relevant from the flood of incoming information. The senses continued to play their own part in discrimination. But more and more the brain became active in shaping the creature's inner reality. That reality grew in elaboration and sophistication. And gradually, from the brain's increased complexity, were born the first seeds of emotion, of creativity, of awareness.

Whereas reptiles came, as it were, prewired, the early mammals evolved to be flexible. That was part of their adaptation. The brain, as it enlarged, took over as chief processor from the senses. And yet, unlike a sense organ, it could be reprogrammed. Especially during infancy, the mammalian brain was malleable; it could learn, as it grew, from experience. And even as an adult the mammal retained some plasticity, so that it could respond subtly, in a myriad of different ways, to its shifting environment.

***

For almost 200 million years the dinosaurs and the other great reptiles reigned supreme. But then, quite suddenly, there came a strange, still dawn when none of these beasts was left alive. All, within a mere heartbeat of geological time, had become extinct. And in their place was a vast ecological vacuum.

Unknowingly, the shy creature that rustled unobtrusively in the forest by night had inherited the Earth. And now it grew more bold. Threatened no longer by fearsome predators, it ventured into the daylight. Ten million years on, it had blossomed into a dazzling multiplicity of forms. Slipping into the sea, it grew to be a giant – a whale. On land it moved into every niche that the dinosaurs had once occupied, then exploited some novel ones of its own.

From a few fragile seeds – those small, harassed insectivores – sprang all of the richly varied mammalian groups. The cats, the bats, the marsupials, the improbable elephants, the impossible anteaters – and one group that was to develop, and go on developing, in a most extraordinary way. This was the group that adapted to the rigors of a treetop life and quickly conquered the digestive problems of eating fruit, nuts, berries, and leaves. A day mover; its vision became keen. Selection favored sight over smell, so that its snout was short, its eyes large and well spaced. Most crucially, in common with all mammals, this arboreal creature's vision, together with its other senses, became more and more strongly brain-centered. Never again would there be a retreat to the eye-centered visual processing of the reptiles.

The tree dweller's eyes began to glide forward to the front of its face, giving a partial stereo view. And the hands, with fingernails in place of claws, were already good graspers. These were the creature's special adaptations, for safe branch walking and leaping, for picking and feeding among the trees.

From this ancient lemur-like animal, over the next 20 million years, evolved the first true monkeys. The thumb and fingers of the hand became better opposed. The nose shrank further. And the eyes completed their migration to the front so that now a broad sweep of the forward vista was thrown into 3-D relief. Such were the overt marks of the simian. And yet there was one other major anatomical change, not so obvious because it was internal – hidden away behind those big, binocular eyes. In fact, it was the most spectacular adaptation of all – a brain of unprecedented size. Never before had there been a creature with so high a brain-to-body-mass ratio. But why? Why did the monkey need so large a brain?

There was no easy answer. (Neither was there for that other group of very brainy creatures, the dolphins and whales.) True, the monkey's diet was varied. But so, too, was that of herring gull or a catfish, and these were not cerebral giants. A monkey's environment was complex. But so also was a crab's, a mouse's, a butterfly's. Monkeys tended to be social. But were they any more so than wolves? Or ants? Taken point by point, there was nothing in the monkey's ecological repertoire that seemed so extraordinary. And yet here was this most extraordinary brain.

Apparently the monkey did not become physically dovetailed into its surroundings. Rather, it developed behavioral pliancy, moving into a kind of "superniche," the occupancy rules of which demanded flexibility of response to the environment. Such a superniche became available only after all the lower niches had been filled. It came last because it built upon everything else, the tip of biological complexity. The prime adaptation for this niche of niches was a talent for adaptation itself. That, in turn, called for considerable mental powers and a large brain to subtend them.

The monkey's ancestors happened to be in the right place at the right time and, most important, with all of the prerequisites for the superniche. Their brains were enlarging anyway to house fine motor-control circuitry for the hands and an improved vision center, while at the same time communication between these two areas were being enhanced by means of a massive neural switchboard. That integration was vital to enable precise hand-to-eye coordination, to allow the finger to accurately close upon and investigate small objects.

There was stimulus, too, for the learning centers of the brain to expand. Full mastery of those dexterous simian hands could come only by experience. Youngsters became good graspers, talented arboreal gymnasts, through learning. (Poor learners tended to fall and be killed or maimed, so were rendered unfit to pass on their "faulty" genes.) The launch pad for a large brain was already there. That is what made it propitious for intelligence, packaged as monkeys, to take off as a survival trait. Instead of relying on brawn, the monkey could use its bigger, better brain to outwit predator and prey.

But that was not the end of it. In time the simian line split. One branch led on directly to the Old World monkeys; the other careered off in quite a different direction. This second group of primates took to brachiating – swinging by their arms – rather than running through the trees on all fours. Their tails disappeared, they grew larger, and their brains grew proportionately larger still. They had become apes.

***

Yet even as the first apes appeared on Earth the global climate machine began to falter. Over the next 10 to 15 million years, average temperatures everywhere tumbled several degrees. The polar ice caps swelled, robbing the world's oceans of up to 200 feet of their depth. There was less rain. And, as a result, in some places where once there had been dense rain forest, the trees began to give way to more open grassland.

This was so in parts of the landmass that one day would be called Africa. Across the Great Rift Valley the chill winds of change blew. Broadening the swathes of savannah there. Driving back the eastern forests to the fringes of Lake Rudolf and the river Omo and other such places. And offering to the apes in these regions a stark choice: Remain in their depleted forest homeland or strike out into a new, exposed territory of wider horizons.

Not consciously, suddenly, or even uniquely was that choice made. There was never some momentous day when a Napoleon of apes marched out onto the plain as if to say, "Now I am man!" It was gradual, tentative, the transition from ape to protohuman. At several different sites it happened, scattered along thousands of miles of the Rift system, from the northeast of the continent to the veldt in the south. Only slowly, imperceptibly, was there a separation between those apes that, unknowingly, rejected the challenge and those that began to make occasional forays onto the savannah.

***

It was a place of small lakes and marshes, skirted by woodland whose trees steadily thinned until they gave way to the grassy plain beyond. In the far distance the land rose, sedately at first, then with more urgency, until it met the azure sky in two great, flat peaks – the summits of long-extinct volcanoes. On their moisture-laden slopes the jungle and its inhabitants still thrived as they had done for eons. But down below, where the rain had become scarcer and the savannah encroached, a new battle for survival was being fought.

Two almost identical tribes were competing for the same dwindling food supply. Much like pygmy chimps they were in appearance and behavior. Members of both tribes were tool users. They brandished sticks when alarmed; at other times they delicately dipped slender straws into termite mounds to lure out the tasty residents.

Both tribes were closely knit, and that was important to them. A lone ape was vulnerable, easy prey for a heavily armed carnivore. Yet there were other reasons for the forging of strong social ties. An individual took over ten years to mature. And in that time it had to learn, through emulation and play, the skills it would need as an adult. The very complexity of the apes' society demanded a lengthy period of indoctrination.

A dominance hierarchy, albeit a loose one, operated within each tribe so that each member had a certain social status. Since every ape knew its place and role, the tribe as a whole maintained a high degree of cohesion and security. But how did individuals know and make known their status? Not by perpetual brawling, that would only have wasted energy and weakened the group. The key was communication. Sophisticated facial and bodily gestures, a range of vocalizations, and intimate ritual acts like grooming, all served to pass information continually back and forth between tribe members. A large part of what a youngster had to learn was concerned with this – the ability to correctly interpret the signals of others and express feelings and moods using the same signaling system.

Thus was the stage set. The two rival tribes spent their days in capricious foraging among the lakeside woods. Leaves, fruits, nuts, roots, insects, small reptiles and mammals, all were grist for their mill. But of late there had been more urgency in the search for food. Hunger, a disturbing new sensation, had made itself occasionally felt in the apes' daily life. And now, in unconscious reaction to this, one of the tribes had begun slightly to alter its habits.

A few days before, a band of young adults from this tribe had cornered an adolescent baboon that had strayed too far from its own clan. Even with its vicious canines, it was no match for the quick-witted apes wielding their heavy clubs. That day the hunters dined well on fresh red meat. And so encouraged, they set out the next day – and trapped an antelope calf.

More often than not the ape hunters were unsuccessful. Their methods were clumsy and haphazard. And the bulk of their food, as for the rest of the tribe, came from more dependable foraging. Yet the hunting continued, intermittently at first, but with each new success fueling the urge to try again.

So, haltingly, the hunts became more profitable. And more sustained. Often now a hunting party of the larger males would be gone from dawn till dusk. They had become bold enough, at times, to break cover of the outlying trees altogether to carry out a raid upon the savannah itself.

That was a most remarkable step, one that no other animal on the planet could have taken so abruptly and survived. And the fact that it was not easy was crucial, for it forced this already talented creature to use its resources to the full.

***

No longer were the apes of the more adventurous tribe indistinguishable from their rivals who had remained behind by the lake. The males, at least, had grown bigger. Bigger because size and strength are important to a hunter, so that selection had increased their size. And though in truth they were only marginally larger, they seemed by comparison to be giants. For now they walked for much of the time, not on all fours as an ape, but on their hind legs, as a stooping, waddling, man-ape. That was not easy for them: Their pelvic structure was still wrong; they lacked the proper bipedal muscles. But the rewards were colossal. They could see form horizon to horizon. And, at last, they had fully freed those marvelous hands.

Now many things began to happen in parallel. There was a gradual division of labor between the adult males, who comprised the hunting group, and the females and their young, who remained generally closer to a home base at the edge of the plain and foraged. What the hunters could not devour at the site of their kill, they brought back for the others. So was the tendency for food sharing, already evident, taken much further. And with hunting and food sharing came also the beginnings of pair-bonding and of stable family groups.

There were times, too, when all the members of the tribe travelled together, foraging as they went. But ever the divide between the sexes grew wider. And it was emphasized still more as an outcome of the changing anatomy of the female.

To walk efficiently upright, the female needed a smaller pelvis, and yet the pelvic girdle could not be so compact as to prevent a baby's head from passing through the birth canal. The result was a compromise. The female's pelvis remained wider than the male's. At the same time the young were ejected sooner, essentially as embryos, before their heads had grown too big. Here evolution had contrived an especially devious plan. For by bringing the baby into the world earlier, it allowed for a greater period of brain growth and of subsequent learning outside the womb. And in turn the young now needed this extra period to acquire the increasingly elaborate skills – communication, tool using, hunting – that a successful hominid had to have.

***

Graybeard never knew what had become of his mate. The tribe had been drinking by the great river when last he saw her. That was some days ago. And now hunger – always, gnawing hunger – had driven her from his mind.

It had been another poor day; the only trophy, a desiccated, fly-ridden wildebeest corpse upon which the cats had already dined. Armed with rocks and bone clubs, Graybeard and four other adult males had managed to drive off some snarling jackals from the kill. But what gristle remained was barely worth the effort.

Picking the last of the marrow from a vertebra, Graybeard watched in the distance 10,000 tons of meat on the hoof that would never be his. A herd of zebra was grazing half a mile away across the heat-hazed plain. For their part they showed little interest in the man-apes, who were normally too slow to be any threat. But Graybeard's gaze had already focused upon a weaker specimen. Fifty yards from the main herd a mare was limping, carrying an injured foreleg. With excited gestures and guttural noises, Graybeard signaled to the others, snatched his cudgel, and led the party out toward the struggling beast. It saw them, of course, and bolted, faster than they despite its handicap. And yet, driven by hunger, by desperate determination, the man-apes would not relent. They knew how to pace themselves, how to choose the speed that was least economical for their prey. Running in bursts, resting briefly in between, they harassed the fleeter-footed zebra until it came to an exhausted standstill. At last, surrounded by these fierce dwarves with their stinging weapons, the mare succumbed.

There was gorging that night in Graybeard's clan as there had not been for weeks. Young and old, male and female alike feasted upon the hunks of raw flesh brought back by the hunters. For now there was contentment – until dawn, when the struggle would begin anew.

They were not yet accomplished hunters. Nor were they yet great tool users. Most of the time, when they were not gathering roots and berries in shrubbery by the lake, or robbing birds' nests, or chasing after insects and lizards, the man-apes were on the bleak savannah scavenging. Meat was their great prize, but the majority of protein-laden ungulates lay far beyond their speed and skill to catch. If they could scare a nervous cheetah or a band of hyenas from a fresh kill, they would do so. But more often than not it was competition with the vultures and jackals for putrefying carrion that they faced.

When they did hunt it was for the immature, the sick, the helpless. Their weapons were still essentially what nature provided them with: stones, for hurling, bludgeons of wood and bone, horn daggers, jawbone saws. Not much. And yet in their armory the man-apes held another tool that, although invisible, was the most powerful of all: a wonderful ability to communicate and cooperate, unique upon the Earth.

Through subtle gestures of face and hand the early hominids could express far more than raw emotion. They could exchange information on practical matters, as well as about their private fears and desires. And they had begun to broaden their vocal repertoire, so that if they made a certain cry, or grunt, or chatter, it would be taken by others to mean some specific thing, an object or action, of importance to the tribe. Still crude and disconnected were these utterances. But 10,000 generations on they would be more refined. For the good communicator stood best to prosper in this heady new world of the hunter-gatherer, and there was strong selection pressure against those who could not express themselves so well.

The same mental apparatus that made for a good communicator tended also to produce a better thinker for other tasks, like devising hunting strategies or new tools. And through these inventions of the improved mind the owner and its offspring were placed at a survival advantage. More advanced tools and hunting methods and more complex social interactions demanded a longer period of learning. So youngsters were dependent on their parents to a still greater extent. And so did the family and tribal bonds grow stronger and more intricate.

Never before was there evolutionary change so rapid and complex as that which assailed the ancestor of humankind. Never before a creature upon which so much selection pressure was focused in so brief a time on one organ – the brain. In a dizzying spiral, the mental prowess of hominids soared. It was as if the universe had labored painstakingly for billions of years to reach some threshold point after which, with reckless abandon, it could hurtle away to infinity.

Through the dispassionate selection forces of nature had the human brain evolved – and evolved in a quite extraordinary way. Already much enlarged in the simians and apes, the twin hemispheres of the neocortex had grown with even greater exuberance along the hominid line. But more surprising still, those hemispheres, so similar in outward appearance, had assumed entirely different modes of neural processing. It was as if the mentality inherited from the apes and earlier ancestors had become condensed and further refined in just one half of the brain, while the other half had evolved with uncanny speed to handle a new, peculiarly human form of thought.

Vital to the hominids' survival was that they begin to peek behind nature's scenes, to understand better the mechanics of their world, to probe and learn, to improvise and create, not complacently accept what was given. So did the notion of objects and events and the relationships between them, however contrived, take on increasing importance. Without realizing it, the hominids became rationalists and technologists. More, they became linguists – inventors not only of tools but of symbols for every new salient concept in the hominid setting. Through symbols, aspects of the perceived world could be held and manipulated in the mind. And since they were symbols only, without substance, they could be toyed with and combined and recombined at will. So was the capacity for imagination and vision brought to full fruition in man. And so also was born the divided brain.

Only within the left hemisphere did these suddenly gained new talents for abstraction and logic and language find residence. True, the right brain kept apace in size and elaboration with its partner. But the right brain never learned how to speak or to reason sequentially. It remained the holistic mind, the synthesist, while the left became the analyst. That was the great divide. Now there were two brains, joined by bundles of connecting tissue, but two distinct brains nonetheless, each of which offered its own quite different version of reality.

For its visual-spatial ability alone, the right brain was indispensable. But it was inevitable also now, for most practical purposes that the consciousness, the personal identity of man, would dwell in the left hemisphere. Alingual, passive, subdued, unfocused, the right brain became the subconscious mind, the poet, the mystic. From now on its presence would be felt only when the left brain was quiet – at night, or in meditation, or in brief, unpredictable moments of inspiration.

Man the engineer, the scientist, the mathematician was in the ascendancy. And the tools of man, and the language, the agriculture, the first towns and cities, the practical skills by which men and women won mastery over their surroundings, were developed in large part through the workings of the rational mind. Not exclusively that way, for transcendent leaps of faith were often the precursors of great breakthroughs. The right brain was ever there, in the background, and its contribution was surely immense. But it was through the logic of the left that the granite foundations of civilization were laid.

***

Fourteen billion years it had taken nature to reach this crucial point. To evolve to the stage where it could begin to analyze and understand itself. And still, mentally, it was an infant. Still its newly developed awareness penetrated no further than the daily thoughts of a Paleolithic man or woman or child.

And could how it have been otherwise? The left hemisphere of the brain had evolved in the first place because its abilities helped the individual hominid (and its genes) to survive. The left brain was not the benefactor of a galaxy ten million light-years away, or a woolly mammoth, or the third moon of Jupiter, or any other part of the cosmos but man. So of course each human being came to think of itself as a separate, self-conscious individual, apart from the rest of the cosmos.

To begin with, as mankind clawed its way out of the Stone Age, there was little thought of anything but day-to-day existence. All higher intellect was focused on matters of immediate practical concern.

Taming fire. Working bronze and iron. Herding animals. Building shelters. Planting crops.

And yet, even in mastering these basic skills, man learned that the world at large was not always so chaotic and violent and unpredictable as it had seemed. That there were patterns and relationships in nature. That the movements of the sun and the moon, for instance, were surprisingly regular. That there was a cycle of seasons, a right time for sowing, a right time for harvesting. And as people gradually won sufficient control over their affairs to begin living in small, stable settlements and, later, in city-states, so they also gained more time for deeper contemplation.

In Mesopotamia and Egypt and Greece there arose cultures in which a small intelligentsia, subtended by the working masses, could spend their adult lives in abstract, pioneering thought. Exploring the physical reality around them. Questioning the nature of matter and space and time. Or developing that most peculiar form of purposive thought – mathematics.

So, over millennia, did civilizations rise and fall. Knowledge was lost and regained and challenged. And ever the mind of man probed more deeply nature's inner recesses.

Came the sixteenth century and the Renaissance movement in Europe. And now, in a newly charged atmosphere of enlightenment, the first true scientists began their work. Among them, Galileo Galilei. He it was who fostered the notion of laboratory experiments, of controlled investigations in which some small part of the world was excised from its surroundings to allow it the freedom to behave as simply as possible. Thus did Galileo put to the test a belief widely held by students of natural philosophy throughout previous ages. That at the heart of all complexity was simplicity. That behind the stupendous, overt intricacy of the universe lay a group of compact, elegant, physical laws.

In truth, it went even further than that. For now mankind stood on the brink of a most astonishing discovery. From his experiments with falling bodies, Galileo found this: The time taken to drop a given distance was exactly proportional to the square root of that distance. No one, in their wildest dreams, had nurtured such a suspicion. That a physical process, an ordinary, everyday happening, might be governed by a mathematical formula.

Apparently there was a link, a subtle but powerful link, between the abstract world of mathematics and the actual physical universe. In some strange way a certain mathematical relationship gave a shadow performance of the drama enacted by the world of experience.

And more was to come. Soon it became clear that not only falling bodies were accompanied by mathematical ghosts of themselves. The genius of Isaac Newton uncovered more equations, stipulating how objects move under force-free and constant-force conditions. He also wrote down a formula for the gravitational pull between any two masses. And, in his explorations of gravity, Newton made another momentous breakthrough. Not only terrestrial things, he showed, were subject to this invisible attraction, but so were objects far away – the moon, the sun, the planets, and, by inference, every other piece of cosmic matter. The physical laws operating here on Earth, it seemed, applied throughout the rest of the universe. Moreover, those laws, it was beginning to emerge, could be expressed in pure mathematical form. Reality everywhere, everywhen, was mirrored somehow in the strange, logical symbolism concocted by man's left brain. But what did it all mean?

***

By the late nineteenth century it seemed, momentarily, that humanity had all of physical knowledge within its grasp. The universe, it looked increasingly, was no more than a giant clockwork mechanism. Set against a backdrop of absolute space and time, the cosmic assemblage of matter went about its business in a perfectly orderly, perfectly deterministic way. If the position and velocity of every material particle in the universe were measured, it was confidently believed, then the arrangement of matter at any future time could (at least in principle) by calculated. In this wholly unpredictable, sterile, mechanistic cosmos, the role of man appeared less and less consequential. Humans were being reduced, step by step, to mere bystanders, spectators of the vast cosmic machine.

But then the unthinkable happened. Hairline cracks began to show in the great edifice of classical physics. Cracks that quickly spread and widened until the entire deterministic view of the universe, first propounded by Newton, split apart and fell spectacularly into ruin.

Why had it happened? Why this sudden demise of Newtonian doctrine? Because mankind had begun to probe regions of reality outside those of normal, everyday experience. Regions of very high speed, very high energy, very strong gravitational fields. And regions, too, more important – of the very small, of the subatomic. Was it so

86

surprising, in retrospect, that the same parochial laws that seemed to operate within the realm of direct human sensation should prove hopelessly invalid beyond that realm? Of course not. No more so than to find in a foreign land that different customs and language prevailed.

Always, before, mankind had experienced matter in bulk. Quarks and electrons broke in upon human awareness, not individually or in small groups, but en masse, fashioned ornately into flowers, or tables, or the ink of this page. But now, as matter began to be dissected, its inner substance teased apart, it changed. It unfolded as something new and unexpected, just as the cells of a flower petal are unexpected, or the molecules within a single cell.

No longer could a subatomic particle be thought of as if it were some wild aberration of an everyday object – a sort of inconceivably small billiard ball. For the fact was that a quark, or an electron, or any other elemental particle was something very different, very alien. The world of the ultra-small was a looking-glass world indeed, but one infinitely more fantastic than any dreamed of by Alice.

How, then, could scientists even begin to describe this newfound microcosm? Words were for human worlds, labels for the inner constructs of the mind. But no such mental model existed for the quark, or the electron, or the neutrino. Here was a domain of existence the nature and behavior of which seemed utterly bizarre.

For instance: At the smallest scale, matter and energy were freely, routinely, interchanged. So that particles could swap their identity (providing they followed certain rules) or even be created out of immaterial, radiant energy.

For instance: In the subatomic world, matter could give up or receive energy only piecemeal, in discrete packets known as quanta. Hence the title of the new rule book for the universe of the very small – quantum mechanics.

For instance: The very nature of quantum reality appeared ill-defined. What at times might act as a solid particle at other times masqueraded as a wave, and a most unusual one at that. For this "matter wave" was not tangible as an ocean breaker or as a ripple of sound but was altogether more elusive – a wave of probability. In some curious sense a quantum particle was smeared out both in space and time. Instead of resembling a hard nugget of matter whose energy and speed and position could always be known, to any desired accuracy, a quantum particle shimmered. Its edges were softened and blurred. The strange wave associated with it supplied a measure of the probability of ambushing the particle at any given time and location. But, disturbingly, a subatomic particle was not duty-bound to be anywhere, anywhen. Simply, there was a better chance of it being found where its quantum wave was more intense.

Beyond visualization, beyond commonplace understanding was this new subatomic domain. And yet it could be described in other terms. In terms of mathematics. More than anything the quantum world was a mathematical world; the quantum wave, a mathematical wave. The arcane equations that choreographed the dance of electrons in an atom, or the outcome of a subatomic collision, were symbol plays of the mind. And, almost incredibly, what these abstract equations of quantum mechanics predicted dutifully took place in the real world. It was uncanny. Here was this beautiful mathematical structure, a product of pure intellect. And suddenly, behold, it had imprinted itself upon the real cosmos, become the actual script that the fundamental particles in nature followed.

Extraordinary.

But was it really so relevant? All these new, exotic notions dealt with such fantastically small flecks of matter and energy. What possible bearing could they have upon the concrete, practical world of man?

Just this: The whole perceived universe, the macrocosm in which humans lived out their apparently well-ordered lives, was fashioned upon a quivering quantum framework. The human world itself was quantal in nature! And the fact that men and women were not personally conscious of the strange goings-on of their subatomic infrastructure was beside the point. Humans saw, because they were only equipped to see, the smoothed, aggregate behavior of countless trillions of ghostly quantum particles and their interactions. That aggregate behavior had been the province of "classical" physics – common sense physics by any other name. But classical physics was a sham, a mirage, albeit a convincing one. At every scale the universe was quantal, now, at the dawn of the twentieth century, as it had been since the earliest moments of time.

***

And so, at last, the universe began to understand itself. Over billions of years portions of it had evolved into tiny, wondrous complexes of matter – humans brains – that could now dimly perceive the awesome scheme of nature and their own role within that scheme.

The purpose of life, the origin and destiny of everything, the existence of some supreme creator. With these momentous issues had the mind of man striven, even as individual men and women went about their daily, humdrum lives. Why were they here? Here on this small ball of rock, circling an inconspicuous star, in an average galaxy, one among billions. Being born, growing, rebelling briefly, spiritedly, in adolescence against dull orthodoxy, before bending inevitably to society's will. Taking up some terribly secure, meaningless drudge work that consumed the best of each day. To earn enough to raise a family. To do the same thing over and over again. Little sparks of joy, pain, hope, confusion lasting – seventy, eighty years? While the planets wheeled, and the stars burned endlessly, enigmatically, in the depths of the night. And the great cosmic saga went on and on. In the face of all this, what was a human life? Small wonder the appeal of a god with whom all such disquieting mysteries could be left for safe resolution. A god who took notice of humanity and watched over it, despite its pathetic triviality amid ten billion trillion suns. A god who asked no questions – and expected none asked in return.

But now, slowly, mankind was waking to a greater realization. To the startling central message of quantum mechanics. That humans were not tragic irrelevancies in the cosmos. Not helpless, petty spectators as Newtonian dogma had long insisted. On the contrary ...

At the heart of quantum mechanics was the astonishing assertion that nothing existed independently of the observer. Nothing existed "out there" until it had been consciously observed. Until it had been registered by an intelligent mind. In quantum parlance, the very act of observation caused the wave function of an object – its mathematical potentiality – to "collapse." For that instant, and only for that instant, of collapse did a particle become real. Once the observation had ceased, the electron, or whatever-it-was, melted back into the mathematical ground of possible reality.

Nor was all this idle speculation or opinion. Experiments with subatomic particles showed increasingly, beyond any reasonable doubt, that it was true. That observership was a prerequisite of the material universe. That there had to be self-aware minds – watchers who knew they were watching – in order to select this reality from all those that were mathematically feasible. Quite simply, a particle was held in abeyance in some curious symbolic form until it was given substance by the act of observation.

And, again, it was not just some peculiarity of the microcosm that could be ignored until it went away. Ultimately, all things – galaxies and trees, and human minds themselves – were composed of subatomic particles, so that this crafting of reality by observation applied to every object in the universe. Indeed, to the universe as a whole.

It seemed outrageous. Compared with the cosmos, the totality of space and time and all it contained, the human brain was so absurdly small and new. How could it possibly have played such a central part in shaping what was real? And yet, in experiment after experiment, this strange truth now shone through: It was meaningless to speak of the reality of a quantum system, and hence of any physical system, until it was actually observed. In some profound way, humans, as self-aware, cognizant beings, participated in the choice of what existed outside themselves. And this applied not only locally. Not only in staged laboratory tests where, for example, beams of electrons were fired through slits at a screen and from the results the effect of the observer deduced. It applied, equally well, over vast reaches of space and time, with breath-taking implications. For in observing the light of faraway galaxies and quasars, it appeared now, scientists were not seeing objects that had already been there, waiting, as it were, for someone to notice them. The very act of recording those ancient photons summoned them into existence, and thus also the remote star cities from which they came. Unbelievable, preposterous as it seemed, the participatory universe was fast becoming fact.

Now mankind found itself no longer at the forgotten periphery of nature, watching passively forces beyond its control. The mind of man was being drawn quickly in toward the creative hub of the universe. Was struggling with the mega concept that, after all, it had some remarkable part to play. And, at the same time, it began to notice that the cosmos around it seemed strangely contrived. That there were coincidences in nature without which the existence of intelligent life – of any life at all – would have been impossible.

Only minor differences in the laws of nature and in the fundamental constants of nature, it now became clear, would have sufficed to make the cosmos uninhabitable. Had gravity been slightly stronger than it was, for example, the universe would have imploded after only a few thousand or a few million years, long before the first stars had had a chance to form. Likewise, even the most modest change to the strong nuclear force would have ruled out any prospect of matter becoming more and more complex, no matter how long the cosmos endured. A slight weakening of the nuclear glue would have caused deuterium to be unstable. And since deuterium was crucial to the hydrogen fusion reactions of all long-lived stars, this would have precluded the formation of the sun and of any life-bearing solar worlds. By the same token, a trivial strengthening of the strong nuclear force would so have sped up the creation of deuterium and of helium-4 in the early universe that, conceivably, no hydrogen would have remained. No hydrogen to kindle slowly in sun-like stars, to steep a world in temperate light for billions of years; no hydrogen for water, for organic molecules, for life.

And it was the same now wherever man looked. A tiny shift of this force or that constant, a trifling change in some basic physical rule, would, it seemed, have so upset the balance of nature that no life could have emerged.

Suddenly, inescapably, humanity stood in the glare of two brilliant spotlights. That of quantum mechanics, with its strange insistence that reality sprang from observership. And that of the new, anthropic cosmology, with its complementary message: the existence of intelligence was somehow directly linked to the structure and state of the universe.

It was a remarkable turnaround. For so long scientists had thought of themselves as the meek onlookers to nature's drama. Purposely, they had tried to interfere as little as possible with the subject of their experiments, to be the outsiders peering in. But now, unexpectedly, unnervingly, from the very heart of the microcosm and of the macrocosm speared these two brilliant beams. Illuminating mankind. Surprising and exposing it. Making it feel uneasy, disoriented, not a little afraid, like a theater audience that has suddenly found itself bathed in light and under the scrutiny of the amused players on stage. So it was that physicists and astronomers of the late twentieth century found the universe looking intently at them. Looking back as if to say, "You are what is truly important. You are the reason and the cause of it all that is here."

**

To begin with, at least, it was too much to grasp. Mankind was not ready to acknowledge this newly revealed cosmic role. Few individuals were even aware of the recent discoveries in quantum mechanics and cosmology that augured such a stupendous change of perspective. And, in any case, such ideas seemed irrelevant to the workaday world, to the frantic daily tasks in the factory or the office or the home. Even more so for those millions struggling just to stay alive.

But slowly, very slowly, awareness would dawn. Already a handful of scientists and philosophers had begun to sense that a great change was coming. And even amid the conservative mainstream of research there was excited talk of uniting all the forces and particles in nature in some grand "Theory of Everything," an all-encompassing mathematical model of the universe.

At the same time man yearned to explore, not just intellectually, but physically, beyond the Earth. So that as soon as it was able, it began to build spacecraft to fly above the atmosphere, to see what could never be seen from the ground. To glimpse the universe in ultraviolet and X-rays and gamma rays. To explore the moon. In time, to visit the more distant worlds of Mars and Venus and Mercury and those, remoter still, in the outer parts of the solar system.

Thus, eventually, was the Voyager project conceived. Two identical probes that in the early 1980s would swing by Jupiter and Saturn to unlock the secrets of these giant worlds. Probes that, once their planetary tasks were done, would travel on and on, beyond the realm of Sol, beyond humanity's knowledge, into interstellar space. Perhaps, after many millions of years, to be found by some other intelligence – by beings who would be curious to know of its origins and of the creatures that had sent the probe.

Remote, extraordinarily remote, was that prospect of alien encounter. And yet it was there. A chance, however slim, for mankind to make itself known to other races dwelling near other suns. And so were the plans made to fix a phonograph record to each Voyager and to protectively coat that record. In gold. In gold that had come together and been deposited over a billion and half years earlier in the great Witwatersrand goldfield of South Africa. And among the precious few ounces of it that were used on the spacecraft was the very atom that bore our particle hero...

A particle that now would help preserve Voyager's interstellar message, "The Sounds of Earth." Help to preserve, in audio-coded form: 117 pictures of the Earth and its inhabitants – images of islands and fallen leaves, of the Taj Mahal, a mountaineer, a sunset, a radio telescope, greetings in fifty-four different human languages and one cetacean tongue (that of the humpback whale); a panorama of Earth sounds, including that of a fumarole, of surf, of laughter, of a kiss.

And, lastly, most poignantly, an hour and a half of the world's greatest music. Music ... for an intelligent race somewhere among the star?... From man?

On a parochial level perhaps it was that. Simply that. Humans trying, by their as yet limited means, to contact other sentient life. Casting from their island Earth an interstellar message in a bottle.

But, on the grander scale, it was more, infinitely more. For Voyager's music had been conceived and played and recorded, in truth, by none other than the universe itself. The universe it was that prepared to send out Voyager with its golden disk. And the universe it was that ultimately would receive it. In Deep Time. A message from the cosmos to the cosmos. In pictures. In words. In music.

Music that had taken 14 billion years to compose. Out of the searing fire of genesis, through the slow forging of galaxies and stars and life-bearing worlds and self-reflective minds. Music that within its ninety-minute program seemed to embrace all of this mystery, this fascinating interconnectedness of reality. For although on one level each piece told merely some small human story, captured some human experience or emotion, on quite a different level the same sequence of notes told a tale of cosmic import.

There was a Bach prelude. A subtle introductory piece. But was it simply the foreword to other orchestral delights? Or did it hint at the imminent unfolding of the universe.

And so to the concert proper. Beginning with a Beethoven symphony. An elaborate composition in four movements. Echoing the birth phase of the cosmos, in which a quartet of fundamental forces made its appearance, one force at a time. And in which all the particles of nature, like the players in a symphony orchestra, participated.

Followed by a Javanese piece, "Kinds of Flowers," the name a reference to two sorts of flowers symbolic in Javanese Hinduism of certain moods. But symbolic, too, pursuing another interpretation, of the blooming of two crucial forms of organized matter – stars and galaxies – in the early universe.

Then "Morning Star," an Australian aborigine song telling of impending dawn, of the awakening of life. Fitting accompaniment to a chapter of the cosmos in which the sun and Earth were formed, in which life subsequently emerged and evolved. Until it became intelligent enough, self-aware enough, to write songs like "Morning Star" and to build a spacecraft to ferry such music to other suns. And to comprehend words and self-reflective thoughts like these.

And then?

To the future. Farther down the river of time that would carry Voyager to the outer planets and, eventually, beyond, to the stars. Evoking images of this, the ancient Chinese ch'in "Flowing Streams," another piece etched on Voyager's disk. Originally part of a larger composition, Towering Mountains and Flowing Streams. Steep mountains down which bubbling streams coursed, to become rivers, to join the sea, to make clouds in the warm sunlight, to cascade as rain on the mountaintops ...

And so to the farthest future. Of a universe that grows forever, and in which matter and energy spread and thin and spread and wind down, wind down. Until there is nothing left but an unutterably dilute mist of subatomic particles in a space vast beyond imagining. Like the sad, stark universe of Blind Willie Johnson, who played the blues. And whose mournful hymn, "Dark Was the Night," was now to be borne starward by the Voyager probe. Into that darkest, bleakest of nights.

Or perhaps not. Perhaps into a very different remote future Voyager would fly. Into that of a cosmos whose growth steadily slows and slows. Until, eventually, it reverses. So that after – how long? – space begins to shrink. And everything in the universe comes back together again. And there is a refocusing of matter and energy. A new crazed, pagan orgy in which the cosmos dances itself to death. Or spectacularly back into life. A scene of savage fury such as that envisaged by Stravinsky in his "Sacrificial Dance" from The Rite of Spring. One more musical offering carried by Voyager into the abyss of Deep Time.

And then?

Then the resolution, the stunning climax of it all. The answer to the mystery of life and being. The Fugue in C. In C Major. Or in creation. A Bach fugue. How very appropriate. Bach, the most mathematical of composers and the most subtle. Lover of hidden meaning, of theme and variation, of endless cycles.

In eight stages it might be summed up. An octave of chapters. Each chapter named for some piece on Voyager's record. A record bearing a quark whose mythical story was being told. While in the background, mind, matter, and mathematics evolved in company as the story moved through Deep Time. As in Bach's music, so in the universe at large, would these ingredients be seen to meld to form reality.

Chapter 5: Flowing Streams

We shall not cease from exploration
And the end of our exploring
Will be to arrive where we started
And know the place for the first time.
—T. S. Eliot

So the terrestrial sojourn of our proton draws to a close. For 4.6 billion years the particle had lain within the Earth's crust, within the core of a gold atom. Until intelligence came, plucked that atom, along with trillions of others, out of the ground and used it in the gossamer coating of a photograph record. And then fixed this strange disk to the side of a spaceship.

10:29 A.M., August 20, 1977: Atop a gleaming, silver-and-white Titan-Centaur launch vehicle sits Voyager 2 – the tip of an arrow pointed at the stars. And now quivering and slowly, steadily lifting as brilliant flame pours from its base, beginning a brief but Herculean struggle to break forever its gravitational bonds with Earth. A thunderous, crackling roar puts a thousand seabirds to flight as the slender missile rises, faster, faster, through the still, humid air of a Florida morning.

Minutes later blue sky has given way to purple, to black, as Voyager reaches the edge of space, its powerful main boosters already spent and discarded. Now the upper-stage motor blazes soundlessly into life, propelling the spacecraft to still greater speeds, out of Earth orbit, toward the chill depths of the solar system, and far away from the home of man. That last, vital surge of acceleration over, Voyager emerges from its aerodynamic cowl like some curious fragile insect shrugging off its pupal shell. Safely in the vacuum between worlds, Voyager no longer has need of its streamlined casing. It unfurls its spindly booms and antennae as an emergent butterfly might stretch out its new wings.

Ten hours later it slips silently past the moon.

***

Never before was there a probe with such a mission: To fly close by the icy giants Jupiter and Saturn, to learn what it can of these great planets and their moons and rings. And, if all fares well, to journey on to Uranus and then even to Neptune, to look upon the previously unseen faces of these far-flung members of the sun's kingdom.

More capable, more flexible is Voyager even than its venturesome predecessors, Pioneers 10 and 11 – the first of the Jupiter probes, dispatched but a few years earlier. Well can it claim to be the consummate invention of early atomic man. An expansive craft by robotic standards, its structural centerpiece, a bowl-shaped antenna – the crucial radio ear and mouth of Voyager – twelve feet in diameter. Nestling below the bowl, a tank of precious hydrazine fuel, and arrayed about this a ten-sided frame bearing Voyager's electronic mind, its communications complex, its tiny thrusters for maneuvering, and its hopeful, wide-eyed message for the stars.

Out from the decagonal frame sprout the craft's lengthier appendages: its magnetometer boom, tall as a two-story house; a pair of antennae like long knitting needles for sampling the sparse fluid of the interplanetary sea; a stockier arm clutching the optical instruments, including two television cameras, and, at the end of another latticework limb, the nuclear-powered batteries that will breathe electrical life into Voyager for the lonely decades to come.

***

By November, Voyager is past the orbit of Mars. By early December it is at the threshold of the asteroid belt, a broad stretch of reefs and shoals beyond Mars that once had seemed as if it might jeopardize all travel to the outer solar system. And yet there is no real hazard to the spacecraft here. Though the asteroids number many thousand, and the elite of these measure tens or hundreds of miles across, yet they are very thinly scattered. Like the twin Pioneers before it, Voyager runs the gauntlet of the flying mountains unscathed.

Less than a year after entering the asteroid belt, Voyager pulls clear of the nether edge and begins to draw near to the first objective its creators have set for it. Jupiter, the superplanet, the star-that-might-have-been, lies directly ahead. No longer a mere point of light but a yellow-orange ball, it swells daily, its four major moons white pinheads in mobile attendance.

Now the probe is only 30 million miles from its rendezvous. Obeying a signal from Earth that took forty minutes to overtake it, Voyager opens its cameras eyes, soon to gaze upon the Jovian realm with a clarity and resolution forever denied to any Earth-bound telescope.

Even as its leading sister craft, Voyager 1, departs from Jupiter safely on course for Saturn, so Voyager 2 begins its own encounter. With a world that is now impressively large. With a world whose "surface" changes kaleidoscopically hour by hour, a live event, because it is no true surface at all but merely the uppermost layer of a formidable atmosphere. Far below that monstrous gassy shroud lies a turgid ocean of liquid hydrogen, untold thousands of miles deep, lit now and then by stark bolts of lightning. And beneath that Stygian sea, only there begins the real surface of Jupiter, the outer boundary of its oppressed rock-metal core.

July 9, 1979: Past the giant, pockmarked moon Callisto sweeps Voyager, moving in toward the parent planet on a wide, arcing path. Sights all around it: a surfeit of moons, large and small; a narrow, dusty ring girdling Jupiter, newly found. And, preeminent, awesome Jupiter itself.

Now by Ganymede, biggest satellite of all, at closest approach only 37,000 miles away. And still inward. Cutting past Europa. Enigmatic Europa, smoothest world of Sol, its icy coat enlaced by veins dark and light. But beneath the ice – a warm, water ocean? Exotic, submarine life? Mysteries enough for many a future probe.

Now beyond Europa and to the point of minimum separation with Jupiter, about 440,000 miles from the topmost clouds. Not spectacularly close – at less than half that distance did Voyager 1 fly by. Pioneer 11 had ventured to within just 27,000 miles. And yet the path of Voyager 2 is well chosen. On this heading it will best supplement what its sister probe has recently learned, while still gaining the gravitational assist it needs from Jupiter to be retargeted toward Saturn. As if hurled by an invisible sling, Voyager corners sharply around the giant planet, gathers speed in the process, then begins its long coast toward its next major encounter.

***

Away from the environs of Jupiter, the awful emptiness of space reasserts itself. Stars, planets, moons – these are seen to be the exceptions, rare jewels in a cavernous, near-empty void. Earth-like worlds are especially scarce. What focusings of matter there are in the universe harbor mainly hydrogen and helium. And even these are extraordinarily far apart.

Only by slow stages will Voyager come to know the cosmic loneliness. The inner solar system is a congested place: a star, four planets, three moons, and various itinerant debris, all confined within a spherical volume less than 300 million miles across. A sphere triple that size encompasses the orbits of the main asteroids of Jupiter, yet by cosmic norms it is incredibly cluttered.

But now the gulf between matter-islands has begun to widen alarmingly. Almost twice as far from the sun as Jupiter does Saturn circle. And, save for a few stray asteroids and comets, there is nothing in between. Steadily the sun shrinks, its rays deprived of their warmth. The dim specks that were the inner planets wink out, and even Jupiter's yellow disk shrivels to become once more just a bright spot of light. Above, below, fore and aft, only the stars remain unchanged, emphasizing their remoteness, whispering a little more clearly the terrible truth of interstellar distance. And beyond that, intergalactic, intercluster. . .

Soon it would be for man to experience personally this ultimate, soul-stealing rapture. For men and women on journeys to the far planets to float free of their spaceships and be surrounded by – everything. To see it without the comforting womblike presence of their home world, any world, near at hand. But to realize, too, in time, that though physical space may be huge beyond deranged terrestrial imagining, it is no more so than the inner space of the mind.

Even now the tendrils of human awareness are reaching out toward Saturn. Stretching across 850 million miles to touch the most stunningly beautiful world of Sol: misty-yellow, many-mooned, many-ringed. And so, at last, comes Voyager, swooping in through the marches of Saturn, after its two-year-long haul from Jupiter. Across the orbit of Titan, mysterious moon obscured by organic orange smog, across the paths of Rhea and Enceladus and Tethys it goes, instruments ablaze. Cameras all agape, pivoting this way and that – the vicarious electronic eyes of the human race at the end of radio nerves almost a billion miles long. For twenty-five months had Voyager slept on its interplanetary trek from Jupiter. But now suddenly it is alert and frantically active again, racing through preprogrammed thoughts in its crude silicon cortex, slavishly absorbing and sifting data and then transmitting these to its masters on Earth.

Did ever an explorer look upon such sights? To Voyager's port: the glorious rings of Saturn, no longer blurred and diffuse but exquisitely, subtlely defined. Rings made of thousands upon thousands of concentric ringlets, differing in size and brightness; sand, ochre, and tan; some braided, others crossed by dark, moving spokes. And within the rings, Saturn itself, its great honey-pale globe less featured than Jupiter's, yet storm-tossed still and the playground for 1,000-mile-an-hour winds.

Again, Voyager's flight corridor is well chosen. On August 25, 1981, the spacecraft comes within 63,000 miles of the Saturnian cloud tops, its speed, timing, and position perfectly orchestrated to allow it to be boosted gravitationally once more toward its next planetary encounter.

And so Voyager enters an uncharted land. No spacecraft before has visited any world beyond Saturn. Moreover, Voyager was designed for only a four-year, two-planet mission. And now clearly the little probe is beginning to shown signs of age. A memory cell in one of its six onboard computers has failed. More such lapses might threaten Voyager's ability to recall instructions beamed from Earth, inflicting it with a kind of robotic Alzheimer's disease. Yet of much greater concern is the apparent seizure of the spacecraft's scan platform. Upon this are mounted Voyager's camera eyes. And upon these are mounted humanity's only hope for close-up views of Uranus for many years to come.

In their laboratories on Earth, Voyager's keepers anxiously seek ways to repair the ailing probe. Like surgeons, those scientists and engineers rehearse over and over again the procedures they will need to carry out on their faraway planet. Yet it is not just a matter of restoring Voyager to its former good health. If the spacecraft is to be as successful at Uranus as it was at Jupiter and Saturn, then it must be improved – from over a billion miles away. And, amazingly enough, that is possible. Because Voyager can be reprogrammed, reeducated, taught to function in new and more efficient ways.

So, as the months and years go, and Voyager draws ever nearer to Uranus, its abilities and intelligence are gradually enhanced. Most crucially, its scan platform is repaired.

But the old probe, too, learns some new tricks. Learns, for example, how to package picture information more economically, a talent that will prove invaluable. For as its distance from Earth widens, the spacecraft must transmit data at lower and lower rates in order that its signal may be discerned from the crackle of background noise. Like a caller straining to be heard over a bad long-distance phone line, Voyager must enunciate its messages more slowly as its distance grows if it is to heard at all.

Nor can Voyager expect any help from home during its upcoming encounter. At Uranus radio messages (traveling at light-speed) from Earth would take over two and half hours to catch up with the spacecraft. So only after a wait of more than five hours could mission controllers be sure that Voyager had properly obeyed a direct command. Yet because of the unique tilt of the Uranian system, the entire close-encounter period will span only a quarter of a day. Through some ancient act of vandalism – a collision with another world? – Uranus has been bludgeoned over onto its side, so that now its spin axis lies almost in the plane of its orbit. As a result, to the approaching Voyager, the dark, newly discovered rings of Uranus and the orbits of its moons will be arrayed vertically. Will resemble, in fact, the concentric circles of an archery target, with the planet itself at the bull's-eye. Voyager is a dart, hurtling toward that target at 40,000 miles per hour.

***

January 24, 1986: The day of encounter. All that Voyager has been set to do, the details of the measurements it must take, the photos it must gather, is implanted in its computer-brain. And now, for the next six intense hours, the probe will watch and listen and monitor as it glides quickly by this unexplored, primeval land.

Dominating the scene by its sheer bulk is Uranus. Aquamarine Uranus, enigmatic, almost featureless, a high haze of methane imparting to it its characteristic blue-green coloration, but at the same time effectively screening from view any turmoil in the atmosphere below. Yet if the planet itself proves disappointingly bland, then its company of moons could hardly be more enthralling. As Voyager spears through the moon/ring plane it gazes upon some of the most bizarre terrain ever seen. None of the Uranian satellites, the largest barely 1,000 miles across, was expected to be other than a stillborn orb of ice and rock. How could worlds as small as these, formed so remotely from the sun, ever have been geologically active? And yet here, in defiance of that:

Titania, biggest of the fifteen moons of Uranus (ten of these discovered by Voyager), riven by huge fault systems and canyons. Evidence of an active youth etched unmistakably upon it.

Ariel, 700 miles wide. Scarred by numerous fault valleys and floes of methane ice. Most of its ancient cratered surface erased by melting. A moon, beyond doubt, that in its first billion years or so was marked by incessant geological upheaval.

And Miranda, a mere 310 miles in diameter. Presumed inert. Yet, in fact, a world whose landscape has been so devastated, so savagely abused, that it may be that Miranda was blasted apart on several separate occasions by impact with rogue asteroids, able each time to stubbornly reform from the fragments of its former self. The starkest canyons on any known world, their walls rising steeply to twice the height of Everest, slice through its surface. And in places it seems as if a huge rake had been brutally dragged across the little moon.

But even as these astounding images speed across the near two-billion-mile-wide void to perplex scientists on Earth, Voyager moves on beyond Uranus. Penetrating ever deeper into space. Heading for its last scheduled port of call.

***

So far is Neptune from the sun that no Earth-bound telescope has ever revealed it as more than a meager blue disk. Its two known attendant moons (more have been found in recent years), large Triton and diminutive Nereid, from afar are faint pinpricks of light. Yet, alien, frigid, and gloomy as Neptune may have seemed, nonetheless it represents now the final outpost of all that is familiar to Voyager and its creators. The decision is taken, over a year after the Uranian encounter, to make this a very special farewell. Commands are issued to the spacecraft to fire its thrusters so as to bring it onto a new and spectacular flyby trajectory.

Onward presses Voyager. Into a region where the temperature is only a few tens of degrees above absolute zero. Where the sun appears 30 times smaller than it does from Earth. And 900 times fainter. So faint that to capture images of the worlds within this realm, the spacecraft's cameras will have to remain trained on their target for many seconds with their shutters wide open. Midday here is as dark as a moonlit terrestrial night.

And now, on August 24, 1989, three and a half years after leaving Uranus, and nearly three billion miles from home, Voyager plunges into the Neptunian realm. Closing from the south, it sweeps through the planet's equatorial plane, skirting the dusty edge of Neptune's broken ring system. And then heads on to the point of closest approach. Barely 3,000 miles above Neptune's north pole, Voyager rushes past, the planet's banded, weather-mottled face relinquishing many a long-held secret to the probe's eager, staring eyes.

And then, just as suddenly, just as swiftly as it had loomed, so the eighth world of the sun retreats and shrinks into the distance. Five hours later Voyager passes within 30,000 miles of Triton before hurtling on, beyond all hope of recall, terribly alone, into the darkening night.

To reach the ninth planet, tiny Pluto, and its outsized moon, Charon, Voyager would have had to do the impossible – to summon all of Neptune's gravitational potential. Only by flying directly through the core of Neptune could the probe have had its path bent sufficiently for it to consummate its grand tour of the outermost worlds. As it is, a quartet of planetary visits must suffice. For Voyager there will no more encounters. No more strange, virgin worlds to explore. None, at least, that will be recorded in human history.

Voyager has run out of planets. But not yet out of luck. Or power. Its generators, fueled by the steady decay of radioactive plutonium, continue to pump thermoelectricity through the spacecraft's vital organs, albeit at a reduced rate. Its instruments remain functional. And so, twelve years into its flight, Voyager embarks on a new phase of its missions, one that will take it to the very brink of the interstellar void.

***

Sunlight that began its journey four and a half hours ago shines weakly on the probe's golden disk. Hurriedly now, Voyager is moving down and away from the plane of the planets. Neptune and Uranus, Saturn and Jupiter are fast-fading points; the inner planets long since vanished from sight. Even the sun, though still many times brighter than any other star in the sky, is now revealed clearly as no more than a star. At this distance its pale rays are shorn of their old intensity and warmth, while all around the alien stars by the thousand crowd. Their ancient patterns – Orion, Cygnus, Centaurus, and the rest – are familiar enough. And yet, in this new, worldless setting, they seem also strangely unfamiliar.

Such is the speed and distance of Voyager now that its native star no longer can lay claim to it. The spacecraft is leaving the solar system for good, and even if every drop of its forty pounds or so of remaining propellant were squandered in a bid to slow its escape, it would be in vain. Its departure parameters are irrevocably set, the spacecraft's fate sealed.

Nor is there much of interest in this ill-lit frontierland between interplanetary and interstellar space to while away the lonely years. The colorful carnival of solar worlds has been left far behind. The swirling, streaming, fall hues of Jupiter. Volcanic, sulfur-festooned Io. Saturn, the Beautiful. Miranda, the Incredible. Uranus, Europa, Titan, Triton, Neptune, and the others, all are gone, shrunken to pebbles, to dust motes. To memories. What remains now is merely this static, hypnotic backdrop of stars and fathomless space.

So, slowly, do Voyager's life currents ebb, its signals weaken. Monotonous and faint have become its murmurings since that final triumphant encounter with Neptune. And yet, even now, Voyager has not lost its penchant for surprise. Ten years after the Neptunian rendezvous and a strange, subtle effect is beginning to make itself felt on the craft. Voyager's position and speed no longer quite match the values that had been forecast for them. Tiny discrepancies are starting to show between the spacecraft's predicted movements, based on the gravitational pull of the sun and the nine known planets, and those that are actually being measured by scientists on Earth. Something else, apparently, some other massive body in the outskirts of the solar system, is tugging on the little probe.

For many years astronomers had conjectured thus. That there might be an unknown object circling the sun beyond Neptune. Ever since the discovery of Uranus and Neptune it seemed that perturbations in the movement of these planets demanded the existence of a still more remote world – the infamous Planet X. Then, in 1930, Pluto was found at just the place where Planet X had been expected to lie. The position was perfect. And yet everything else, curiously, was wrong. The newfound ninth world turned out to be a planetary midget, smaller even than Earth's moon, and much too light to have any appreciable effect on the orbits of its giant inner neighbors.

So the hunt for a massive perturbing object had gone on, through into the 1980s and early 1990s. Until now, at last, in 1998, it seemed the quarry had been tracked down. As analysis of Voyager's aberrant motions gradually revealed, there was indeed a previously unseen mass far out in the solar system. By 1999 enough was known of the mystery body and its position to begin a search for it with ground-based telescopes.

And, suddenly, there it was: A tiny white speck, moving slowly, slowly, against the background stars in the southern constellation of Corvus. Four times as massive as Earth. Moving in an orbit almost perpendicular to that of its sisters, ten billion miles out, taking 800 years to circle once around the sun. Planet X! Or, as it became christened in the year 2000, Prospero, tenth and outermost world of Sol.

Not for another two centuries would scientists recognize the full irony of that name. Shakespeare's Prospero it had been who, by magic, raised a shipwrecking storm off the island he inhabited in The Tempest. And this newfound world was indeed rightly suspected of having caused great storms on Earth by occasionally diverting the paths of remote comets so that these icy missiles rained down upon the inner solar system, wreaking havoc. Infrequent but catastrophic collisions with such objects had brought about the sudden extinction of many species on Voyager's home planet, among them the dinosaurs. In fact, it was as a result of Prospero's comet-hurling antics that the mammals, and, eventually, man (and Voyager), had been able to evolve so spectacularly.

But what was not suspected, not for many years, was that it had also been Prospero that had blundered into Uranus four billion years ago. Prospero that had knocked Uranus onto its side, causing the seventh planet to cast off a whirling disk of rock and ice that had in turn coalesced to form the Uranian moons and rings. And so – the irony – it was Prospero, just as in Shakespeare's tale, that had fathered enigmatic Miranda.

Well before this Uranian link was realized, though, Prospero had tempted human curiosity unbearably. A world whose surface had never been seen, about which virtually nothing was known, but a world that may have had a profound effect on life's evolution – it cried out for investigation. And so, very quickly, a robot probe was built to fly close past it. A probe in some ways like Voyager, but technically three

106

decades more advanced. Driven by nuclear electric propulsion and managed autonomously by computers of enormous power and intelligence, the Prospero probe set sail in November 2004. Just six years later, cruising at a speed of 130,000 miles per hour, it surpassed Voyager's distance from sun.

And so, even as Voyager saw a new generation of spacecraft begin to overtake it, the old explorer reached the end of its active life. In 2011, Voyager crossed the heliopause, the magnetic boundary where charged particles from the sun ram into the surrounding gas of interstellar space. Feeble indeed were the signals it then sent back to Earth, seven billion miles away, of this, its final discovery. Its nuclear power source was nearly exhausted. Four years later the last trickle of data from Voyager entered a tracking station in the Mohave Desert. And then . . . silence.

Legendary now became Voyager's exploits. To future spacefarers they would rank alongside those of Sputnik, Apollo 11, Columbia, and Mayflower II, the first of the lunar colonizing ships. And when men and women had settled beneath the pink skies of Mars, and on the chill plains of Ganymede and Triton, they would recall to their children how a small, primitively equipped probe had blazed a trail across the outer solar system for others to follow. Yet where Voyager had gone to now, no one could say.

***

The final message to Earth had flashed by in February 2016, a last attempt to renew communication with the veteran craft. But Voyager had already stopped listening.

That had been 147 centuries ago.

And now some . . . unusual changes have taken place around Voyager's native star. Mankind has remodeled several nearby worlds, dismantled half a dozen moons for some disturbingly large-scale projects, and begun recently tinkering even with the energy-making processes of the sun. There are new modes of travel bordering on the magical. For up to interplanetary range – the transfer booth. An individual steps into one, perhaps in Madrid, dials a destination, and is reconstituted half a second later in Jakarta, or Atlanta, or forty minutes later (though it seemed instantaneous) in a hotel lobby on Callisto. And for long-distance journeys, of course, there are the starships.

Starships five miles long. With propulsion systems that allow them to cheat Einsteinian barriers by tunneling between widely separated points outside of conventional spacetime. Which is why the trip to Alpha Centauri now takes only nine hours. Why the exotic fruits of Procyon IV can be delivered fresh to the eating houses of Sol. And why man is already on a thousand worlds, and swiftly moving out farther, exploring, adapting, evolving.

And then there is Voyager, unnoticed star flotsam. Forgotten relic from the earliest space-faring days of the human race. Now lazily, mindlessly spinning, as it had been for over 14,000 years. Drifting at eight miles per second, a speed almost comically low set against the tens of trillions of miles in which average distances between neighboring stars are reckoned.

Voyager was not built with interstellar flight in mind. And even now, after all these centuries of travel, it has covered rather less than one light-year, so that it remains closer still to the sun than to any other star. Only now, in fact, has it come to the limit of the sun's dominion – to the Oort Cloud, vast store house of frozen cometary nuclei encircling the sun four and a half trillion miles out. Slow and tiny amid the traffic of starliners, Voyager has the prominence of an ant crawling down a freeway.

And yet, in another sense, the old probe far exceeds any of these latter-day spacecraft. For it has become immortal. It no longer ages. Away from the matter-strewn environs of stars there is nothing to erode metal or glass. Nothing save perhaps an occasional wayward dust particle. So, as if enclosed within a timeless vacuum, Voyager wanders inexorably on. Toward the stars.

But which stars?

Even the scientists who launched Voyager, millennia ago, had a partial answer to that. They knew exactly which way and how fast Voyager was moving as it left the solar system. And they knew, too, roughly how the positions of various stars in the sun's neighborhood would change over the next few hundred thousand years. So it was simple enough to compute the approximate times and distances of Voyager's earliest stellar fly-bys.

Easy enough to predict that the closest Voyager would probably come to another star within half a million years would be 1.65 light-years. Nine trillion miles! And that this almost unnoteworthy "rendezvous" would be with an obscure M-type red dwarf known simply as Ross 248. Indeed, there was nothing in those long-ago interstellar forecasts to quicken the pulse. Nothing to suggest that

Voyager might come within striking range of extrasolar worlds. Its presence be detected by some sentient species fortuitously living upon them. Its message from humanity discovered, decoded, answered. . .

In fact, around 40,000 AD, when Voyager made its closest pass to Ross 248, the only intelligence nearby that star was mankind itself. Long ago the robot probes from Earth had arrived here to scrutinize every planet and moon and asteroid in the little star's domain. To be followed by man in person.

And it was the same at other neighboring stars by which Voyager flew. Ross 154, DM +15 3364, Lalande 21185, Sirius. Everywhere that Voyager went the human race had long since gone before it to colonize or simply to explore. If ever the craft's humble message had been found within the first million years, it would have been by the extra-terrestrial progeny of those who had written it.

But eventually the habits of man changed and the material traces of humanity were no longer so evident among the stars that Voyager encountered. To what new heights mankind had risen was yet hard to say. But the sights of human-made cities and spacecraft, reassuring despite their futuristic appearance, had suddenly ended. And, again, Voyager was alone. More alone then before.

Even the constellations had become distorted, until now they were unrecognizable. Partly because of Voyager's slowly shifting viewpoint, but chiefly because of the stars' own motions, the Great Hunter and Ursa Major, Perseus and all of the other heroic asterisms had dissolved. In their place were fresh patterns. Alien patterns.

***

And so in time Voyager came to new reaches of the galaxy. After tens of millions of years it wandered into the realm of an old orange supergiant and its tiny, white dwarf companion. So vast and tenuous were the larger star's cool outer layers that occasionally they wafted out as far as the dwarf. And then the smaller sun would greedily gather up some of the billowing gas around it. Gather up more and more, until suddenly, when it could hold no more, it ignited its newly won atmosphere in a flash of fusion fire. Once every few centuries, as Voyager floated by, the dwarf would erupt spectacularly in one of these brilliant nova outbursts before quickly fading again. So to the passing spacecraft it seemed like a bright point of light, winking on and off, at the edge of an orange haze.

And there were other such intriguing encounters. One with a pulsar – a ten-mile-wide, furiously spinning ball of pure neutrons. Spewing out light and charged particles in two narrow, sweeping beams, lighthouse style. To within a billion miles Voyager came of this crushed star, remnant of some old supernova blast, to meander briefly among its pathetic collection of charred worlds.

Giants and dwarfs, stars alone and in groups, suns with planets and without, the spacecraft looked upon. Through monstrous, dusty interstellar clouds, rearing like storm heads out of the galactic plane, it flew. Through nebulae both somber and bright, through clutches of sparkling new suns recently formed. Ever on. Millions of years passing. The probe's invisible trail stretching out for thousands of light-years behind it.

And now it has been eight million centuries since Voyager left Earth. The flowing streams of time and gravity have brought the little craft within scope of a modest yellow star, its disk growing in apparent size with each millennium that passes. A trillion miles distant, 100 billion, 10 billion. Rapidly the star's brilliant globe is swelling, its system of planets coming into view, spreading, growing. The sun!

But no . . . not the sun. A star much like it. But slightly smaller, slightly dimmer. And with only seven planets, one of these twice the size of Jupiter.

Along a steep, hyperbolic path Voyager descends into the inner regions of this new solar system. On its present heading it will barely miss the second planet, orbiting 80 million miles out from the parent star. Closing. A white dot. Closing. A blue-white ball. Closing. A blue-white-brown marbled globe of continents, polar caps, great oceans, and swirling clouds. One of those rarest and most precious of cosmic creations, a world rich in liquid water. A second Earth! And there is oxygen in the air. The unmistakable signature of life. . .

But what kind of life? Given similar but not quite identical conditions to those of Earth, what variety might alien life-forms take? Will most of the land animals be quadrupeds? Will there be birds with feathers? Fish with scales and fins? And, if there is high intelligence, must it of necessity stand on two legs and look at its world through two eyes?

On the nightside of New Earth's terminator, scattered twinkling lights shine out of the darkness – and they are not volcanoes. If only the view of them were closer . . .

But swiftly the encounter is over. And whether, in passing so close, Voyager was heard or not by the dwellers of those great cities, it will not bide near this planet long enough to be studied further.

Chapter 6: Dark Was the Night

Now entertain conjecture of a time
When creeping murmur and the poring dark
Fills the wide vessel of the Universe.
—William Shakespeare

Millions of microscopic craters pepper its metal frame. A ragged hole, teeming with stars, gapes in its high-gain antenna. Otherwise, in the year 5 billion AD, Voyager is intact.

Five billion years: In the previous such era human intelligence had synthesized from stardust. In the last one millionth of that term man had progressed from building its first cities to engineering lasers and genes. Now what the human race had become would not even recognize the tiny, battered spaceship as its handiwork.

Man has gone, or changed indescribably. Yet the galaxy – the familiar Milky Way – remains much the same. From still-plentiful gas and dust in the spiral arms, the mightiest of suns continue to coalesce, streak through their bright careers, and erupt sensationally at the end as supernovae. Together with more subtle ways of shedding stellar matter, these titanic outbursts replenish the galaxy for future generations of stars.

At the opposite extreme, red dwarfs, many as ancient as the galaxy itself, glow feebly and interminably. What is another five billion years to these stellar Methuselahs? Their modest hydrogen reserves are for 50 billion – perhaps even 500 billion years – at the miserly rate they are used.

But it is not so with Voyager's native star. Somewhere, lost in the fairy dust of a spiral arm, the old sun is dying. Already it has doubled in size and brightness since Voyager departed. On Earth the climate has altered radically, the polar caps melted. Exotic new life-forms have emerged to suit the changed environment, and yet even these thermophilic species are doomed in the traumas to come.

***

Three billion more years elapse. Like a feather in a whirlpool, Voyager is swept around the galaxy a dozen more times. Now, in 8 billion AD, the sun is swollen grotesquely, has become a red giant, 150 times its original size. Solar luminosity is up 2,000 times, temperature down 1,800 degrees centigrade to a ruddy 4,000 degrees.

And already the old sun has cannibalized two of its attendant planets. First to go was Mercury, its old orbit engulfed by Sol's distending surface layers. Then Venus succumbed. Now Earth is the new hell-world. Oceans boiled dry, its scorched mountains and plains are hot enough to melt lead. And it is utterly, permanently, scoured of life. Only the noisome clouds of gas that vent ceaselessly from its surface proffer any movement.

So it ends as it began. For a second time in its history the Earth is lifeless, searingly hot. A great star illuminates its daytime sky, but this is not the Sol of old – compact, energy-rich, and benign. The monstrous florid sun that now blazes down sprawls across half the sky. Seething, its outer atmosphere reaches to within 30 million miles of Voyager's birthplace.

What has happened since the spacecraft left home? Why has the aging sun changed so profoundly?

First, around 5 billion AD, hydrogen in the sun's core started to sputter and run out. This it was that had kept the sun shining more or less steadily for eons preceding. Now, choked with helium ashes, its useful fuel gone, the old solar fusion power plant was condemned to shut down. With no outward radiation pressure to buoy it up, the sun's core began to sag under its own weight. And in the process some of its stored gravitational energy was set free. Exactly half this energy helped supply radiation from the surface; the other half drove a stellar central heating system that catastrophically altered the sun's appearance. Heat migrating outward from the collapsing core caused the sun's entire atmospheric envelope to mushroom. And with that the galaxy gained a new red giant.

Continuing to shrivel under its own gravity, the sun's central region became hotter and hotter. At the fringes of the core, where hydrogen was still abundant, the temperature climbed to ten million degrees. And at this point a hydrogen-burning shell flared up around

114

the core. Like a superscale brush fire, it began to spread outward through the surrounding layers of the sun. Energy released by this mobile fusion reactor now contributed to the light that escaped from the solar surface.

Such is the sun's condition eight billion years after Voyager's launch. The cosmic clock ticks on . . .

Steadily the sun's sleeping helium core grows hotter. Gravitational contraction pushes the temperature up to 100 million degrees and sets the scene for – helium flash! Suddenly the core is alive again with fusion reactions. Helium nuclei smash together to form the heavier elements carbon and oxygen. And as the helium burns so its carbon-oxygen ashes are dumped upon the shrunken core.

But helium is a poorer source of energy than hydrogen. Also, at the much higher temperature needed to fuse it, the reactions occur faster and the core helium supply is exhausted after only a few million years. Again, the sun's central reactor shuts down. A second shell, this time kindled by helium, fires up at the perimeter of the hot carbon-oxygen core and follows in the path of the still outward-moving hydrogen-burning shell.

Then comes an uncertain phase of development of Sol. By one mechanism or another the sun starts to lose matter at a fairly high rate. Over a few thousand – at most a few million – years, it sheds virtually all of its dilated hydrogen-helium atmosphere. The outer layers may escape piecemeal by way of a gusty stellar wind. Or they may be shot out en masse as a bright expanding shell of gas. Whatever the means of ejection, all that eventually remains of the sun is a white dwarf – a hot, dense, planet-sized ball of inert matter.

***

Voyager travels on. By 20 billion AD, its old home star is a shadowy, crystalline sphere of super-thick carbon. Its planet of origin, still bound by gravitational allegiance to the spent solar dwarf, is dark, dead, frigid as space. The star system that once nurtured the human race has expired.

20 billion AD: Now the cosmos has doubled in size since Voyager set sail. It is bigger because space-time has expanded, become more rarified, somewhat cooler. And yet in character the universe is little changed from the days when some large-brained biped took its first halting steps starward.

And the Milky Way? That, too, looks familiar enough. Still hugging the galaxy's plane are the warm, glowing nurseries of protostars. Elsewhere an occasional supernova bursts, showering several sun's worth of fertile matter back into the galactic melting pot. Stellar reincarnation goes on.

Another 20 billion years pass, and another, and another.

For 100 billion years the universe has endured, Voyager for four fifths of that time. And now there are palpable signs that the galaxies are aging.

In the Milky Way the star-making machinery has ground almost to a halt. What unconsolidated gas remains is too thinly spread ever to claw itself together again by gravity.

Slowly the galaxy changes hues. The white and blue of its once-majestic spiral arms fade as, one by one, the most massive stars disappear and are no longer replaced. Eventually even sunlike stars become rare, leaving only red dwarfs to light up the Milky Way with their dull, ruby glow.

It is the same throughout the rest of the universe. And though maybe a few renegade, late-developing spirals remain tinged with youthful blue, before long these, too, slip into senescence.

Voyager turns one trillion. And now the golden epoch of stars is nearly over. Even the lightest, most venerable of the red dwarfs are coming to the end of their hydrogen reserves. So the soft glow of the galaxy's spiral arms, halo, and core steadily fades. Sometime between 1 trillion and 100 trillion AD the last of the stars in the universe goes out. Every star has consumed its usable fusion energy stock and subsided under self-gravity to become a cold, dark cinder.

Most common in this gloomy far-off age are the relics of small stars. These, the white dwarfs – Earth-sized or smaller – that have chilled and darkened to velvet blackness. In 100 trillion AD nonluminous white dwarfs, or "black dwarfs," populate every corner of the galaxy.

Scarcer are the dense remains of heavier suns – the neutron stars. Scarcer yet, though in numbers much greater than when Voyager began its journey, are the dark, sinister carcasses of the heaviest stars of all – black holes.

Around many of these bygone suns, charred, lifeless planets still wheel. Mercury-like inner worlds may have been vaporized as their parent stars bloated into red gianthood. But the larger fraction of

116

planets, though raped by stellar explosions, survive intact. Will they now go on to circle their old gutted suns forever?

Recall that stars move. They move relative to one another. And although stars are as dust motes floating in a cathedral, they do once in a while brush past one another. Close stellar encounters may be rare. But when they do happen the planetary systems of one or both stars may be thrown into turmoil.

How close must close be? Roughly, a star with a given planet needs to be approached by another star to within the radius of the planetary orbit. When that happens the orbit will be badly disrupted.

And the interval between such encounters? Take the average density of stars in the galaxy to be one per 35 cubic light-years, that is, one sun for each box of space with sides 3.25 light-years long.

Next take a planet, like the Earth, whose orbit is 186 million miles across. Assume the parent star drags the planet through the galaxy at 30 miles per second. Then, by the time the planet's orbit has swept out 35 cubic light-years of space, it will likely have been crossed by an alien sun. With the values used here, this time comes out to be just over seven trillion years.

And so Voyager enters the second great era of the mature universe. The first ended when the last of the stars went out. Now, in this dreary ensuing phase, planets are being freed from their primeval moorings.

How many close encounters does it take to wrench a world completely out of its orbit? And how many encounters are needed to pry loose an entire planetary system?

Voyager floats in a cosmos black as pitch. Endowed with human eyes it would see nothing. Almost nothing – from the heart of the Milky Way come occasional mysterious flashings. And from the depths of other galaxies, too, sparks briefly fly to interrupt the long night.

For the most part the liberation of planets proceeds in the dark. Slowly, painfully slowly, each star in the galaxy is stripped of its worlds. Possibly 100 close encounters are needed to shatter a solar system entirely. In the more rarefied, peripheral reaches of the galaxy, the process may be delayed – there are fewer chances for stars to meet. But in Deep Time all the worlds throughout the cosmos slip their gravitational moorings.

At some point between 1000 trillion and 100,000 trillion AD, the second great era of the future universe draws to a close. Now every world is unbound. Like their decayed parent suns, they steal along private lonely trails through galaxies cold and dark.

And so the third far-off era of the universe begins. The galaxies themselves are poised to evaporate.

***

Envision a star as if it were a molecule in a liquid. Struck hard enough, through collision with its neighbors, and a molecule may escape from the liquid's surface. So, in similar style, may stars break free.

Just as stars can come close enough to eject planets, so they may, on still rarer occasions, approach so close as to hurl themselves free from the gravitational pull of the galaxy. During such a near stellar miss one star gains kinetic energy at the other's expense. And if the boosted sun gathers enough speed from the encounter, it will simply break the shackles of its galactic orbit altogether.

The gravitational "slingshot effect" – that is what the scientists who built Voyager would have called it. Indeed, Voyager itself once used the slingshot to tour Sol's outer planets. As the spacecraft encountered Jupiter, in July 1979, it purposely followed a path that would utilize the Jovian gravity field to hurl it on, at increased speed, toward Saturn. What kinetic energy Voyager gained was at Jupiter's expense, so that the fifth planet was slowed minutely in its orbit and made to fall imperceptibly toward the sun.

In Deep Time dead stars in all the galaxies experience the slingshot effect over and over again. Multiple close encounters banish star after dead star into galactic space. By 1 million trillion AD, nine tenths of the mass of every galaxy has vaporized in this way – planets, dust, and gas, we well as stars – while the remaining one tenth, robbed of kinetic energy by the same near collisions, has been drawn by the galactic gravity embrace into an increasingly dense core.

Perhaps there had always been a supermassive black hole at the Milky Way's heart. If not, it makes little difference now. By the close of this era all major galaxies in space will have acquired such dark, fearfully dense cores. And into one of these great bottomless pits of gravity every unevaporated star remaining is about to plunge.

***

Half familiar is the scene – a strange, distorted echo of the infant cosmos. For were galactic black holes not once active before? As quasars – the brilliant cores of youthful star cities? Those laser-bright beacons of the early universe were born of matter straying too close to the black hearts of galaxies. And now it seems they are primed, ready to burn again in this far-flung era.

Steadily, inexorably, what matter is left in the Milky Way joins a swirling maelstrom around the central core. A trifling fleck deep within this frenzied whirlpool is all that remains of Earth – and even that is about to be incinerated. Of the other solar planets, most were cast out of the galaxy trillions of years ago. The black dwarf of Sol, like Earth, is ensnared in the hot vortex. Hot vortex – and growing hotter. And brighter. Glowing now, the Milky Way's core bursts into brilliant candescence. All around the universe the galaxies are putting on their final show – going quasar once more.

Perilously close Voyager skirts to the outer edge of the black hole's domain. Dead ahead – a straggling neutron star, itself already doomed. White light from the blazing necklace of the black hole dances off Voyager's interstellar record, a message that now will never be played – less than half the original spacecraft remains intact.

And now Voyager is upon the neutron star, banking sharply around its precipitate gravity field. Again the slingshot effect. And suddenly Voyager is reprieved, hurtling outward, away from the central danger zone.

From afar the spacecraft's pitted golden disk, like a glazed eye, stares at the Milky Way as the old galaxy devours itself. Light rays, some testimony to the demise of man's home planet, rebound from "The Sounds of Earth" – then spear farther into Deep Time, drawing out yet another subtle cosmic thread.

So Voyager departs from the galaxy within which it was forged. And travels on into a greater void.

***

The other galaxies in space: What has befallen them? In a far-off time, when the cosmos was new, the star cities crowded close together. Huddled into clusters and superclusters, many of them lay within easy range of human telescopes.

But the universe is awesomely larger. It has grown 100 million times in size while Voyager has been in flight. Compared with the vertiginous gulfs between galaxies now, the Big Bang was but a cosmic stone's throw away when humans lived.

And yet some galaxies – their dark corpses, at least – still cling gravitationally to one another well into the third cosmic era. Maybe the superclusters have not survived the universal expansion; their member clusters have likely been torn away by the continued stretching of space-time. But the individual clusters are much smaller and more tightly bound by gravity.

Still, over trillions of years, it is hard to presage the fate of these minor galactic tribes. Collisions within them will doubtless be many. If the galaxies run in to one another, they may merge. Or, by the slingshot effect on a grand scale, a galaxy may be expelled from its family unit. Smaller systems will tend to be absorbed by more dominant neighbors. It could even be that a cluster will reduce itself to a single supercluster well before the era of mass stellar exodus.

And so the long-term prospects for the Local Group are unsure. If the group survives in any cogent form at all, then by late in the third era – that of galactic evaporation – it will be a queer place indeed.

Exiled from their parent galaxies, dead stars, planets, Voyager, and other flotsam roam free. Of the Local Group galaxies, once thirty-strong and star-bright, the shocking remains are five supermassive black holes, each boasting over 10 billion solar masses crushed within a radius of one light-day.

Like colossal baleen whales, the galactic black holes glide through space, their gravitational maws open wide, swallowing everything in their path. Over hundreds of millions of years they trawl upon whatever scarce plankton they can find. And then, by chance, two of these voracious monsters meet. Out of their bizarre union emerges a still-larger black hole. And after nameless eons more this new, engorged creature melds with another of its rare breed. And so it goes on.

Between 1 million trillion (10^{18}) and 1000 trillion trillion (10^{27}) AD, all the clusters of galaxies in space have reduced themselves by cannibalism to solitary multigalactic black holes. The sole representative of the Local Group, massive as 100 billion suns, spans a light-week. Fading now, its ten-million-degree halo, starlike in the distance, shines wanly on Voyager's ruined frame.

***

Such then is space in the year 10^{27} – an ever-widening ocean in which 90 percent of all cosmic matter wanders free in the form of extinct stars, planets, lesser debris, and individual subatomic matter. Well beyond comprehension now is the extent to which its contents have been diluted. On average, one sun's worth of matter enjoys a privy estate of 10^{28} cubic light-years. One hundred suns in a bubble of space as big as the entire twentieth-century cosmos!

The remaining 10 percent of matter is locked up in a billion or supermassive black holes, caliginous fossils of the old galaxy clusters. Slippery and sheer are the gravitational walls of these abysmal space-time pits. And yet so far apart are the black holes that their influence upon one another is as nothing.

Galaxies without stars, stars without planets, and everywhere the ruin of gravitational collapse. What further devastation can the universe suffer?

Marvel at Voyager's luck. It has a close companion – a black dwarf barely a million light-years away.

If only now Voyager had infrared sight, an eye that could perceive heat waves. Heat waves? In a universe this old, this rundown, this utterly cold?

And yet there is something strange about that paltry black dwarf a million light-years to starboard. Somehow it contrives to stay warm. Not warm in the sense of human fireside comfort. But its temperature, even so, is well above the ambient frigidity of space.

Odd. How could this ex-sun without fusion fire warm its carbon bones? Not by any macroscopic means. So how?

An ancient memory stirs. The breaking of cosmic symmetry in the first microstratum of time . . . the origin of the four fundamental forces . . . grand unified theories (GUTs). Yes, there was a condition – absurdly irrelevant in those first frenetic instants – that these sweeping designs of nature decreed. All baryons must decay. Even that most secure brick of normal matter, the proton, must eventually crumble.

If those grand unified theories – those GUTs – hold good, then the proton is unstable – though what a most durable brand of instability it is. Between one million trillion trillion and one hundred million trillion trillion years might pass before all of the protons in the universe break up, a time span that seemed safe enough when men and women first read of it . . .

But no longer. Now the secret of that tepid black dwarf is laid horribly bare. Quietly, ever so gently, it is being simmered by the decay of its protons. And yet this is no new process. For if protons are unstable, then huge numbers of them have been decaying since creation's dawn. Like corn popping, 1,000 trillion trillion of them would have split apart annually inside old Sol. But while the fusion furnaces roared, any trivial warming through proton decays would have been hopelessly masked.

Now, though, in 10^{27} AD, space is chilled to absolute zero. All conventional ways of generating heat have gone. And the minuscule contribution made by decaying protons is at last becoming apparent.

How then does the disintegration of a proton heat a dead star? The proton simply delivers itself spontaneously of a shower of energy-rich positrons, photons, and neutrinos. Only the neutrinos – light-speed wraiths – escape directly to the stellar surface.

In the claustrophobic interior of a black dwarf, a newborn positron soon confronts its opposite number, an electron. And when that happens, because they are particle and antiparticle, they annihilate each other. Their sole legacy – two identical high-energy photons that, as they are promptly absorbed, serve to heat minutely the surrounding stellar matter.

So, by proton decay, is the surface temperature of a black dwarf raised – to about one degree above absolute zero. And cool as that may seem, it is toasty indeed compared with the wintry chill of the cosmic background.

Nor are black dwarfs the only victims of this sinister warming. The decay of communal neutrons – also demanded by GUTs – acts as an immersion heater for neutron stars. Indeed, because these heavier, more imploded suns are more dense, they are kindled more vigorously by particle decay. Almost a roiling inferno in this frozen future age, the surface temperature of the neutron stars soars to minus 170 degrees centigrade.

And yet it is a costly reprieve from oblivion. Once all the protons and neutrons are gone, at an agonizingly slow rate, there will be no more stars. As their baryonic skeletons crumple every black dwarf and neutron star in space will vanish in a puff of photons and neutrinos.

***

But do protons decay? There was a time, recorded in the dusty chronicles of man, when that question was undecided. And if protons are stable, what would it mean in the course of Deep Time?

No proton decay, no skeletal collapse of black dwarfs and neutron stars. Both these forms of derelict sun would endure well beyond 10^{32} AD. But would they last forever?

Again, it is with the inner, miniature world of matter that the answer lies. Delving into the psychedelic scriptures of quantum mechanics, the relevant chapter is found, its topic a curious one – that of "tunneling."

To recap. In classical physics a particle without the energy to leap over a barrier can never cross to the other side. Plain enough – a tennis ball hurled at a wall could never slip through ghostlike through the stonework . . .

Or so it might seem. But then classical physics is rooted in common sense. And common sense is mere heuristics garnered at the human level. Among the finer layers of nature it no longer holds sway, deferring instead to the looking-glass rules of quantum mechanics.

And what the quantum rule book says is this: Particles are not tiny tennis balls. Even tennis balls are not what they seem, for given enough time one would indeed succeed in passing ghostlike through a solid wall. All matter has a wavelike, probabilistic character. And in the realm of subatomic particles, this fuzzy, wavy behavior is everywhere manifest.

Real particles are waves. They can behave as miniature Houdinis. That is the quantum message. Every particle, however starved of energy, has a finite chance of crossing any barrier. So that even if it cannot scale it, a particle may eventually burrow through a barrier. Hence the term "quantum tunneling."

And there is a familiar enough example of it – ordinary radioactive decay. Strictly forbidden by classical physics, the breakout of an alpha particle (a bound state of two protons and two neutrons) from a radium nucleus, for instance, is pure quantum escapology. The alpha particle summons its wavelike alter ego to find a passage through the high mountains of electric potential ringing the nucleus; it lacks the energy to scale conventionally the mountain peaks.

So, once again, nature proves itself capricious at heart. And yet is this new phenomenon of tunneling really so cavalier; its effects, viewed more generally, so unreasonable? All things tend toward a state of lowest energy. In classical science most subatomic arrangements of matter would be forever frozen in a higher state – a sort of perpetual cosmic tension. Only is it by quantum tunneling that particles may occasionally slip through otherwise insurmountable barriers and so relax.

The effect works not just with nuclei normally considered radioactive – those of radium, thorium, uranium, and so on. Because of quantum tunneling almost every kind of nucleus in the universe is ultimately unstable. The one exception is iron. Iron nuclei lie in the deepest of energy valleys, so that to transmute they would have to be fed with energy from outside. All other nuclei strive to become as laid back as iron. Lighter nuclei, over vast periods, would tend to fuse together spontaneously by tunneling, steadily edging their way down the energy slopes into the "iron gorge." By the same process heavier nuclei would incline to fragment and approach the idyllic ferrous state from the opposite direction. The period required for all this – between 10^{500} and $10^{1,500}$ years. On such a time scale every scrap of ordinary matter in the universe would transmogrify to iron.

$10^{1,500}$ years – and yet even this colossal period seems brief compared with others that quantum physics can yield.

Consider: A black dwarf is a relatively high-energy system. It dreams only of the day when it may turn into a free-and-easy neutron star. And indeed, because of quantum tunneling, it can realize this ambition. The only holdup is the fantastically wide energy barrier that separates a black dwarf from its desired neutron condition. For the dwarf to tunnel through the barrier would take $10^{10^{76}}$ years! In a subsequent period of similar length the neutron star would burrow itself again to the paradisical relaxed state, a black hole.

But . . .

**

In Voyager's true cosmos it is the fourth great future era – that in which all remaining protons perish. Now the very fabric of the wrecked spacecraft is starting to rot and fall apart.

10^{30}, 10^{31}, 10^{32} AD – and suddenly it happens. The proton decays. The proton that emerged from the foaming space-time of genesis. That was buffeted in the storm of a nascent quasar core. That wandered into a new galaxy – the Milky Way. That lived, transiently, in the heart of a supergiant star. That was enshrined in a gold nucleus at the instant the massive sun erupted. That settled into the crust of an embryonic world, to be mined four and a half billion years later by sentient bipeds. That was incorporated in Voyager 2. That . . .

Shower . . .

That decayed.

A positron somewhere – if "somewhere" has meaning still when reference points have gone. A positron born of the proton and shot from the surface of this shapeless, dissolving lump that once had been

Voyager. And now this positron moves unrestrained in a vapid ocean of terrifying dimensions.

Spring in the year 100 million trillion trillion: Supermassive black holes are the only islands, unutterably small and lonely, in a cosmos awash with electrons, positrons, photons, and neutrinos. All "heavy" matter has eroded into these tiny, insubstantial grains. As a whole, the universe has enlarged 10,000 million trillion times since the age of man. And now the average distance between each electron and positron is greater than the diameter of the old Milky Way.

These electrons and positrons: Whence did they come? Not from the old black dwarfs and neutron stars. As the protons inside these objects broke apart, the resulting high-density gruel of particles and antiparticles annihilated itself almost totally. No, the pelagic electrons/positrons that remain came largely from the decay of protons already floating free in the cosmic void – from the archaic interstellar and intergalactic gas.

And now, in nature and scale, the universe has become altogether alien. Even that most mundane event, one subatomic particle bumping into another, has all but ceased – the separation distances are too great. Everything is desperately dull and banal.

Yet the same laws of physics govern the universe as before. There are particles that exert forces. And so those particles continue to interact.

Being opposite in charge, electrons and positrons find each other attractive. They can even form a pair-bond, the result of their strange alliance – positronium, a hydrogen-like atom in which a positron acts as a lightweight proton with the electron circling around it.

At first, even in the year 10^{32}, the positrons and electrons remain too jittery to settle down into bound states. The cosmic background temperature may have sunk as low as 100 trillionth of a degree above absolute zero. But still the energy of motion of positrons and electrons is too high for them to condense out as positronium.

Ten times more the universe ages, ten times more . . . 10^{35} AD, 10^{40}, 10^{50}, 10^{60}. Now the entire age of bright stars seems less than the wink of an eye. More incomprehensibly brief is it in relation to this future cosmos than the quantum gravity era, spanning just 10^{-43} of a second, seemed to humans.

10^{65} AD, 10^{70} . . . and then – change! In roughly the year 100 billion trillion trillion trillion trillion trillion, the positron that came from Voyager finds a mate. With its companion electron it forms a single atom of positronium. Yet never was there an atom such as this. The orbit of the two particles around each other is as wide as the entire twentieth-century universe! Unbelievable as that may seem, this vacuous piece of matter is still millions of times smaller compared with its own huge cosmos than a humble hydrogen atom (less than 10^{-8} of an inch across) was in relation to the universe when humans lived.

Everywhere the same remote marriage of electrons and positrons is taking place. At last their energy of motion has fallen below the attractive energy invested in their bound state. An ethereal sea of positronium rapidly fills all of space.

But even that is not the end. As soon as an electron and positron conjoin they begin spiraling in toward each other. When finally they meet their first kiss proves deadly. Particle and antiparticle, the couple realization too late their fatal difference. In an instant they extinguish one another. And then, in fine theatrical style, their photonic spirits race apart as a pair of empyreal gamma rays.

Not that this self-destruction of the positronium sea is a quick affair. It takes about 10^{45} times as long as the age of the universe when the positronium first formed. So only by 10^{116} AD, have the majority of positrons and electrons contrived to annihilate each other. And well before that a far more striking event has taken place.

***

Recall those other inhabitants of this far future cosmos, the black holes. And now look more intently at one of these seemingly permanent, immutable caverns of space-time. With hypersensitive vision focus on the space around it, see every subtle shift of energy there, the flight of each subatomic particle as if it were a shooting star.

And observe this: A weak though steady stream of particles is flowing outward, away from the unseen gravity pit. Slowly, very slowly, as if by a phantom wind, the black hole is seeping energy into space. Yet how can this be? Had we not believed that a black hole was a prison from which nothing could ever escape?

But then we had not read thoroughly all the pages of that shimmering bible of quantum mechanics. For if we had, we would have come across a passage, strangely scripted, with the title "Virtual Particles." And even by the standards of this weird science, its message would have seemed bizarre: All of space, even an apparent vacuum, is in truth crammed full of particles – electrons, protons, and every other type.

Realize, though, that these virtual particles have negative energy states, and as such they cannot normally be seen. Only when one of them is jolted especially hard is it able to materialize in our world as an ordinary particle with positive, measurable energy. At the same time the gap or hole it leaves behind in the dense virtual ranks appears as an antiparticle.

So how to make a particle-antiparticle pair? That, in fact, is not so hard. Only focus enough energy at a single point and the "pair-creation" process will take place. The problem is, matter and antimatter make poor company. Upon meeting they totally annihilate each other, giving off an intense burst of energy. Normally when a particle and antiparticle flash into existence, the partners remain close together and so mutually destruct in short order. That is an end to the affair. But consider now what might happen in the region around a black hole. Within this fierce gravity zone particle-antiparticle pairs would certainly be made. Yet before the antagonistic partners could recombine and destroy each other, one of them might tumble down the clifflike sides of the black hole, a victim of its intense gravity pull. Its mate, now safe from annihilation, might escape into the outside universe.

And the source of energy to make this brand-new particle? There can be only one possibility for that – the gravity field around the black hole. Steadily the black hole is sapped of its gravitational energy. Which is to say, it loses mass. Black holes evaporate! And although the process is slow at first, it gathers speed as the black hole gets lighter. Toward the end, when the black-hole mass is really low, the process accelerates tremendously. Eventually it becomes explosive and the depleted black hole ejects all the rest of its contents back into the cosmos in a blaze of gamma rays.

So the theory goes. The rate of evaporation depends only on the mass of the black hole. If, during the Big Bang, mini-black holes were made with the mass of pebbles, these would have erupted almost as soon as they were formed. Those as hefty as mountains would have completed their vanishing act around the time Sol was lighting up its nuclear fires. And those, asteroid-like, of between one billion and ten billion tons, would have been the endangered species at the time humans first conceived of them.

Bigger black holes of the stellar variety may take until the year 10^{66} to give up all their spoils. And the supermassive breed? They are the most durable of all. Only by 10^{100} AD will black holes of 100 billion solar masses be spilling their last gamma beams. Still more ponderous black holes, boasting the contents of a whole galactic supercluster, could – if ever they formed – persist until the year 10^{108}.

So the scene on the universe's 10,000 trillion trillion trillion trillion trillion trillion trillion trillionth (10^{100}th) birthday: All protons and neutrons have decayed. Most of space is replete with an incredibly dilute gas of positronium, steadily annihilating itself. Every stellar-size black hole has popped. And those exploding now are the biggest ones of all, harboring the crushed matter of entire galaxies. It is a spectacular finale. The universe ends with a firework display!

Except that the universe never really ends. It goes on, growing bigger, colder, darker, and sparser. Or does it?

**

Spill a cup of coffee on a desk and there is an absurdly slim chance that all of the coffee molecules will hop spontaneously back into the cup again. Granted, the odds are stacked high against it and field tests could prove messy. Yet the future cosmos has time enough to work through its myriad permutations. Along the road to eternity there is ample opportunity for even an ultrathin particle soup to convene, by chance, as a blob of matter, a star, a white rabbit, or anything else.

Here is the ultimate challenge to a universe that expands forever – to regenerate itself, to find a pattern amid a gathering gloom of chaos. But how?

Envision a glass tank with a central divider. In one half is the green gas chlorine. In the other, ordinary air. The divider is removed. Now the molecules of chlorine and of air may freely mingle. Each molecule darts along a random path, rebounding off the sides of the tank and off other molecules. Within minutes the green of chlorine is suffused evenly throughout the tank. And so it would remain, in all probability, for trillions upon countless trillions of years.

In all probability. But every microscopic arrangement of that chlorine-air mix is just as likely as any other. The overwhelming majority of possible states happen to be chaotic ones in which no large-scale pattern stands out. But some of the states are special. In these the chlorine is unusually lumpy. In a few exceptional states all of the chlorine is back neatly in one half of the tank.

More molecules than there are stars in the universe are involved in such an experiment, so the chance of an exceptional state – a predefined state – cropping up is vanishingly small. But still it exists. The spilt coffee could flow back into the cup. All Earth's atmosphere could congregate over Iceland at the stroke of noon. And the weak brew of particles in the future universe could suddenly, alarmingly, rush together to form a new Voyager or Mussolini in a spacesuit. Who is to say that the reader's cosmos is not some 18 billion years old but a freak special arrangement of a universe immeasurably older?

Still, it does seem like wild speculation. Do things in the real universe not habitually wind down? Oases of order shrink steadily as the great surrounding desert of chaos encroaches? There is even that base scientific precept, the second law of thermodynamics. "Use energy for any purpose," this avows, "and some of it will inevitably be wasted, degraded into a less valuable form."

From the second law comes the prediction that a quality called entropy – a gauge of disorderliness – will always rise within a closed system. Order becomes anarchy. The pristine cosmic card deck of the Big Bang is shuffled more and more until, in Deep Time, it is made totally random.

At least that is the way the universe seems to go. And yet, in truth, it is a giant gambling game. All the second thermodynamic law does is affirm at the macro level the fact that in microscopic systems there are vastly more possible chaotic arrangements than orderly ones. Hence the odds favor chaos.

The mighty fortress of the second law rises tall and proud – from shifting sands of probability. When will its foundations collapse, its walls subside and crack? Not on any reasonable timescale perhaps. But in Deep Time?

Shuffle a card deck long enough and it must eventually return to its original perfect suit order. With only fifty-two "particles" to sort, the time required is triflingly small – a scant trillion trillion trillion trillion trillion years or so.

Throughout the whole cosmos there are rather more than fifty-two particles. Each, too, has a somewhat richer choice of locations than have the cards in a deck. So it should take longer for randomized particles in the universe to be shuffled back into any kind of order – much longer. The precise number of years cannot fully be written out –

it would overflow not only these pages but all of twentieth-century space-time.

No. For the rising tide of entropy to turn back cosmos-wide must involve manic time scales. And yet not necessarily infinite ones. Within Deep Time there is room aplenty for any period, however long, that can be contemplated. And that gives unlimited scope for natural experiment. The imagination runs riot:

Eventually there might come another universe identical in every detail to that of the reader's. But not before there had been billions and untold billions of others, similar, but with petty deviancies. And not before there had been still hosts of others that were completely different . . .

A cosmos in which, for instance, every particle was organized in a cubic lattice . . .

Or in which there was only one star . . .

Or in which Voyager collided with Neptune . . .

Or in which . . .

Ad infinitum.

Thus the liberal logic, or illogic, of cosmic perpetuity. Everything is allowed – and everything an infinite number of times. If such were the universe, it would be nightmarish, a crazed, pointless cycle.

As it is . . .

***

A date unknown: The denudation of the last black hole belongs to a forgotten past. Even the final decay of the positronium sea took place lost eons ago. Now there is only blackness, ultimate cold, space without end.

And a positron – the positron.

In company with some other electrons and their antiparticles, it survived the breakup of the positronium sea. And now these last specks of matter, together with greater legions of photons and neutrinos, are all that remain of substance in the universe.

How else could it have ended? With the cosmos rebuilding itself? That was mere armchair theory, a physicist's pipe dream. It never came to anything. Nor ever will it. For what the human mind could not grasp, mercifully, is the utter desolation of this place. Were just two particles to come within a trillion trillion light-years of each other now, it would be a staggering juxtaposition.

And, remember, space-time is still growing. The average gap between particles is widening, not narrowing. So that, as times goes on, the prospect for some impromptu local gathering taking place becomes ever more bleak.

Is this then the true future universe: Space, huge and burgeoning, the home of solitary particles in eternal isolation? All of it cooling, thinning forever?

One positron alone. And only one question unanswered about the fate of everything that surrounds it: Will the cosmos in time fully unwind? Will the energy of all particles in the universe at last be smeared over a random thermal distribution? If so, this would be the "heat death" – the supreme blah – that nineteenth-century physicists first contemplated.

At the moment of heat death, entropy would peak at its maximum possible value, unity. Chaos would rule unchallenged.

And the alternative? That would only happen if photons born of the decay of black holes and of the positronium sea stayed clear of collision with other particles. Then might the photons avoid ever sharing their high-grade energy and thus coming into thermal balance with the rest of the cosmos. In this case the universe would drift closer and closer to the dismal heat-death state but never actually reach it.

But what difference does it make – 100 percent chaos or 99.999 percent chaos? Either way, in time, the universe ends up dark, dreary, hopelessly washed out.

In time. In Deep Time. At the far end of Deep Time, for that is where the cosmos has arrived. And, paradoxically, it seems that time's condition now is no more certain than it was at genesis.

In the beginning there was the problem of time's source: If time had a moment of birth, what was there before that? In the prenatal state of genesis how could anything have happened, preparations been made, outside of time?

And now this new riddle. In the far-future universe change had ground virtually to a halt. With particles so widely spaced, even the dispatch and receipt of a solitary photon becomes almost a divine event. And without change what significance has time?

Perhaps time's axis needs new labels. Did it ever really make good sense to think of the Big Bang as being at Time Zero?

Moving back closer toward the Big Bang, events were crowded much closer together. Within the "first second" there was more change

– a higher density of events – than in all the subsequent history of the universe put together. And edging back ever nearer to the point of creation, events were squeezed still more tightly into thinner and thinner microslices of "conventional" time.

Take then a bold step. Redraw the map of time. Rather than the Big Bang at t = 0, mark it at t = - infinity. Then, looking back, it becomes genesis, not Armageddon, that lies an infinite period away. And the remote future of the cosmos? Put that at Time Zero!

The result – a fresh view of the scheme of things. Rather like those tongue-in-cheek antipodean charts with south at the top and Australia, lording it, big and prominent, just above center.

And, too, the new temporal plan may offer more than just novel perspective. With the Big Bang set at t minus infinity, time stretches out farthest in the direction of most activity. Which is promising. Also, it removes the stubborn problem of "What came first?" since before any given event there would always be an infinite number of others. The quest for a genesis spark would be doomed to frustration from the outset.

As to the future universe, the new scheme literally has no time for it. As events become fewer and farther apart, time loses the framework against which it was previously measured. Beyond a certain point, with every particle effectively cut off from every other, time would cease to have any meaning at all – the point of Time Zero.

Is it wild surmise? Will time really come to an end at some stage in the future universe? Or will it live on, ethereally aloof from the dispersed wreckage of matter, energy, and space? Again, it turns on the true nature of time. And that, as ever, remains elusive.

But one thing does seem clear. Even if time itself survives indefinitely, its direction – the way its "arrow" points – must come increasingly into doubt. Why? Because time's forward direction is taken to be that in which order dissolves into chaos, in which entropy increases. When all that remains is total chaos – chaos changing into equivalent chaos – there is no longer anything by which to distinguish past from future. Even if occasional events still happen, they cannot, even in theory, be assigned to any certain time order.

And so, to the end of everything. A positron, whose fabulous life story has run its course, now hangs in stasis. Entropy has peaked. Confusion abounds. Time is . . . uncertain.

And yet all of this, all of this vast, bleak, moribund gloom into which the universe swells indefinitely, had once been seen as but one possibility in the formative days of scientific man. It was the destiny presaged for an "open" universe, a universe lacking the means to prevent itself from growing forever. Even as Voyager fled its native star system, humans debated between this and a second, equally remarkable, but radically different, cosmic fate.

Chapter 7: Sacrificial Dance

Do not go gentle into that good night,
Old age should burn and rave at close of day;
Rage, rage, against the dying of the light.
—Dylan Thomas

What, asked man, if the universe were not open? What if the universe harbored sufficient matter – in forms both luminous and dark – to slow, and eventually reverse, the ebb tide of genesis? What, in other words, if the universe were closed? Then might the future cosmos take a very different course. Then might there come, in time, not a dark, impenetrable night, not infinite expansion and gloom, but . . .

10:29 AM, August 10, 1977: Atop a gleaming, silver-and-white Titan-Centaur launch vehicle sits Voyager 2 – the tip of an arrow pointed at the stars. An arrow now quivering and slowly, steadily lifting as brilliant flame pours from its base, beginning a brief but Herculean effort to break forever its gravitational bonds with Earth.

Minutes later blue has given way to purple, to black, as Voyager reaches the edge of space, its powerful main boosters already spent and discarded. Now the upper-stage motor blazes soundlessly into life, propelling the spacecraft to still greater speeds, far away from the home of man . . .

Past the moon . . .

Beyond the orbit of Mars . . .

Past giant, pockmarked Callisto . . . by Ganymede . . . beyond Europa, enigmatic Europa . . .

To the point of minimum separation with Jupiter . . .

With Saturn . . .

With Uranus . . . and Neptune . . .

And so, twelve years into its flight, to the very brink of the interstellar void.

Hurriedly now, down and away from the plane of the planets, Voyager is moving. Another decade on, and the weak but just-discernable lure of Prospero begins to make itself felt upon the failing craft.

In 2011, Voyager crosses the heliopause. And then, four years later, falls silent.

Centuries pass. Thousands of centuries.

And ever Voyager drifts on, forgotten. Alone. Toward the stars. Toward remote encounters with Ross 248, Lalande 21185, Sirius, and others in the solar neighborhood. Lazily, mindlessly spinning. Slowly, steadily drifting. Drifting, shifting its vantage point, until even the constellations become twisted and strange. And so, at last, the ancient primitive crafts strays into new reaches of the galaxy. Into the realm of an aging orange supergiant and its nova-prone companion. To within a billion miles of a pulsar in impossible pirouette. Past bejewelled star clusters and towering dust clouds and bright, gassy nebulae, light-decades wide, hauntingly beautiful. And, in time, to the outskirts of a new solar system – to a family of seven unfamiliar planets huddled in the warm glow of a sunlike star.

A trillion miles the range. One hundred billion. Ten billion. Rapidly now the star's brilliant globe is swelling, its brood of primary worlds revealed, spreading, growing.

Along a steep hyperbolic path Voyager descends into the heart of this new stellar domain. On its present heading it will barely miss the second planet.

Closing. A white dot. Closing. A blue-white ball. Closing. A blue-white-umber globe, marbled with continents and oceans, with dazzling, ice-capped poles and swirling clouds.

Closing. Now skimming and bouncing across the upper atmosphere like a flat stone on a pond. Hull temperature rising, rising. But Voyager was not built for reentry!

Ninety miles below, from a balcony in a city of exotic design, a child gazes up with wide, multifaceted eyes. Eyes set in a pale green, Lepidopteran face whose innocence, whose wonderment, even alien features cannot disguise. For thirty seconds a small orange star, flashing erratically, sails silently across the night sky. Not a true star, nor a meteor, nor one of the new orbital power platforms – that much she knew. But then what was it?

Gone.

Scorched and with a three-foot-wide hole newly burned in its high-gain antenna, Voyager plunges once more into the safe vacuum of space. The bright plasma sheath that encased it briefly during its fiery atmospheric transit is dispersed, its metal skin cooled.

And now Voyager travels on. For billions more years it travels on. While space-time expands, and the galaxies age. And the star-making machinery of the Milky Way grinds almost to a halt.

***

And so only slowly, imperceptibly, does this new myth begin to diverge from the old. For eons the outward rush of the clusters of galaxies continues. Even the most generous human estimates of cosmic mass could scarce avoid a lengthy expansion phase. A trillion years might be conservative, the upper limit indefinitely large. So that, inevitably, at least some of the major epochs of any credible model universe, open or closed, would be shared. The exhaustion of stars, escape of planets, and breakup of galaxies – these would as likely befall a closed cosmos as one that was open. Given a longer period of expansion, every effect up to and including the death bursts of supermassive black holes could be encompassed.

If closed, however, the universe would eventually reach some maximum size, after which it would begin to contract. Now new phenomena, not seen in the earlier myth, could be expected to make their appearance.

***

In this new, alternative cosmic future it is the halfway point – the very instant of reversal. Having swollen to its greatest diameter, the closed universe has just begun to deflate, like a gigantic balloon that has sprung a slow leak.

Tens of billions of years pass. And at first it seems almost as if the cosmos is merely falling backward through time, as if the contractive phase were an exact temporal image of the era of expansion. Might it even be so? That, on an Earth-like world, during the contractive cosmic stage, wind and water would conspire to build mountains out of sediment, with rivers flowing uphill? That material, scattered in air and soil, would draw together to form the living bodies of senile creatures that would subsequently grow younger, their life processes running in reverse? An old man would "mature" into a boy, then a newborn baby, and, eventually, an embryo and an egg? And yet, since the workings of the brain also would be time-reversed, an individual's actual perception of events would be no different than it was for the expansion phase of the universe?

And yet suppose – the more likely case – that this was not so. That the closed universe does not retrace precisely its own steps while shrinking. Rejoin the myth sometime after the moment of reversal. And notice something very odd, very unexpected, is starting to happen. Has not the total energy of the universe started to rise? In violation of energy conservation? Then suddenly the reason for that dawns clear. The myriad of photons throughout space – the cosmic background radiation – is now having its wavelength shifted, not toward the red (low-energy) end of the spectrum as before, but toward the blue. As the contraction of the universe gathers pace, photons are being blue-shifted more and more, their waves being compressed rather than drawn apart, thereby made increasingly energetic. So that eventually they surpass the energy with which they were first emitted.

Eons more go by, and a time is reached when the universe has regressed to the size it was when humans first stood upon the moon. Yet it is not the same universe as man knew it then. The temperature of the cosmic background radiation may have returned to three degrees above absolute zero. But the stars are long since exhausted, cold and black.

As space continues to shrivel, so does the temperature of the universe rise and rise. Closer and yet closer the ruined remains of the clusters of galaxies approach one another. And now the cosmos is again the size it was just one billion years after the Big Bang. Crowded are the galaxies once more, the clusters beginning to fuse into a single, vast hypercluster, lodging all the star communities there ever were. Then this one hypercluster itself stars to condense, until the galaxies within it collide and combine. Still the cosmos continues to collapse, shrinking to its ancient size just one million years after the outrush of genesis. Now aged stars are careering crazily toward each other at speeds rivaling that of light itself.

And another extraordinary thing is taking place. Because of the steeply rising blue shift of free photons in space, the night sky is becoming increasingly hot. Around the year 100,000 BD (Before Doomsday) the temperature throughout space reaches that of a normal stellar surface. Burned-out remains of stars – assuming any have survived – are heated up. Phoenix-like, they blaze into light, their remaining nuclear fuel burning at a prodigious rate. Some explode, others are simply vaporized, their constituent particles set free to join a hot consommé that is then squeezed, and squeezed further, by the ever-contracting boundary of space.

Now the universe is withered to a size it knew just one year after its birth. And the greatest of black holes, the lightless cores of extinct galaxies, have begun to feast voraciously. Gobbling up first much of the boiling, 100-million-degree soup around them, unsated still, they devour each other, growing larger and fewer in number. Ultimately only a single black hole commensurate with the shrunken cosmos might remain – a black hole within a black hole, so to speak. But perhaps before that the universe would reverse its direction once again, rebounding from its oppressively crushed state and transforming, at the very last fraction of a second, Armageddon into a brave, new genesis.

To some twentieth-century human thinkers there was philosophical comfort in such a cosmic model that promised rebirth. And advancing the thesis a simple step further, an oscillating scheme for the universe emerged in which all of space-time cycles between expansion and contraction, each regenerating Big Bang the dramatic successor to a devastating Big Crunch. A myth as splendid in its own way as that of the endless night.

***

And yet in all this, one factor, one supremely decisive factor, has been left from the reckoning. The universe is not mindless and unaware. It has intelligence, consciousness. And through that it may have the power to control its own destiny.

Chapter 8: Fugue in C

We are the music-makers, and the dreamers of dreams.
—William Shakespeare

There were no landmarks. No galaxies. No stars. Only a heartbeat ago the universe itself had seemed to erupt, gushing forth all the matter and energy there would ever be. One second old was the universe, and it teemed with the fecund brew of creation.

Out of that brew, in due course, would emerge minds, fashioned of stardust. Minds that presently would learn of their amazing origins. Would begin to see that they were, literally, starchildren, the awesome product of 15 billion years of cosmic evolution – the universe, at last, able to reflect upon itself, to marvel at itself. And they would read these words, conceived in stardust, printed in stardust, with their stardust eyes and brains. And for one exhilarating moment they would forget their small cares and share in the certain knowledge that each particle of which they were made had been forged in the genesis fire, had seen the slow, majestic birth of galaxies and the bright, dawn quasars, and then had ventured on, across eons and light-millennia to meet and fuse in the substance of a human form. Even as other particles had come together, from all reaches of the cosmos, to make sea gulls or solar flares. Or spaceships.

***

And how quickly in the end it had culminated. For even as Voyager 2 fled the Earth in 1977 there were those still alive who could recall the first mechanically powered flight at Kitty Hawk just seventy-four years before. Meteoric though man's biological development had been over the previous few million years, still more spectacular were the social and cultural explosions that followed. Yet these were as nothing compared with the speed of technological change.

With the harnessing of steam in the mid-eighteenth century and the Industrial Revolution this triggered, man's control and dominion over nature had begun to spread at a fabulous, unprecedented rate. Often it seemed that technological change was far outstripping the human capacity to adjust at a social or personal level. There were those who wanted to slow, or stop, or even reverse the dizzying spiral of automation and innovation.

And of course it was clear – frighteningly clear – that not all the new devices of man were being put to benign use. In the thermonuclear bomb, in the range of other exotic weaponry hung in threatening array at the edge of space , in the toxin-gushing factories and cars, and in the dancing, mind-manipulating images of television there was ever perceived the specter of Armageddon. Despite its lofty accomplishments and still loftier ambitions, mankind seemed on the very brink destroying itself and the natural world that had nurtured it.

And yet, just as obviously, man was an intrinsic part of nature. man-made meant nature-made. For whence had the human race come if not from the soil and the sun and, billions of years earlier, from the scattered remnants of more ancient stars? And billions of years before that, from the elementary particle brew of creation? Humans were fashioned of the same raw stuff as the trees and the rocks and the spray of the sea, whether they cared to contemplate it or not. Even their sleekest contrivances, their cold glass-and-steel skyscrapers, whose soaring forms belied the stunted aspirations of those who worked antlike inside them. And their elegantly spun clothes hiding the animal within. And their anodyne shopping malls and push-button, air-conditioned homes. All these pretentious trappings were packaged stardust and sunlight, no less natural, no less perverse for all the fastidious attentions of man.

Of course there was immaturity here. But it had to be seen in cosmic context. Man the technologist was not apart from the universe, a virus that had it had invaded somehow from outside. In every sense man was the universe, learning how to grow, exploring its newfound self-awareness.

As yet that awareness was still dull and highly fragmented, scattered across billions of small, individual minds on Earth and who could guess how many myriad more alien minds throughout the rest of space. At this stage in their growth minds were still atomic, still the property of individuals. Only gradually would they merge to form a more unified and coherent whole, just as only by slow degrees had the separate subatomic particles of the primitive universe come together to make gas clouds and planets and prebiotic soups.

It was to be expected – was inevitable – that these island-minds should have different thoughts and opinions, even as in the early universe particles had darted about independently, helter-skelter, exploring all physical options. Some minds played a full and energetic part in developing the new technology. Others, the majority, were simply swept along by it. And yet others resisted it vehemently, and so also played a crucial role in what was to come. These latter-day Luddites who rebelled against technology often did so because they believed that man was harming nature, ravaging in a few decades the complex, fragile ecosystem that had taken billions of years for the Earth to create. And it was true: The human race was affecting its surroundings, often adversely, more than any other living creature had done before. It was exterminating whole species of animals and plants, polluting the seas, poisoning the atmosphere, cremating the virgin jungles.

But man was the Earth . . . and the universe. And it was no more "unnatural" that men and women should hunt the black rhino to extinction, or tear the heart out of the Brazilian rain forest, than it was for a volcano to erupt. Or for an ape-man to step out onto the savannah. If the Earth was being hurt, then it was hurting itself. And so it would have to learn. Just as a child learns when it scrapes its knee or becomes sick through eating poison berries.

Not only was mankind a child of the universe, it *was* the child-universe. And when, in 1997, it shot dead the last wild black rhino, and when, in 2018, it had turned all of the Amazon Basin into a scorched, sterile wilderness, it learned a hard lesson. The rhino and tens of thousands of other species had died for that lesson, and much of what was held beautiful had gone. But now the universe would not forget. And in time, in Deep Time, it would make good the violence it had done itself.

***

Just twelve years it had taken Voyager to reach Neptune, but already much had happened back on its creators' world. More than ever Earth's fate and that of mankind seemed to hang in the balance. Homo

143

sapiens, self-styled "wise man," held a nuclear shotgun to its head and cried aloud across the continents and oceans it had despoiled, "I do this to protect myself from you!"

And yet, even amid the rubble of such folly, science and technology had continued to bloom. In biochemistry, in robotics, in the physics of semiconductors and superconductors, and in a score of other fields, new possibilities had emerged the mere thought of which would have been ridiculed only half a century before.

At the same time the ranks of the environmentally concerned had swelled. For at last it was becoming transparent to all that much of what humans had done over the previous few centuries had been self-harming. In the end it was not pity or magnanimity to its fellow species that spurred mankind to change its ways but the stark, imminent threat of extermination. Survival of the collective self was the driving force now, as it had been since the dawn of life on Earth, since the dawn of the universe itself.

Above and beyond its parochial planetary mission, Voyager was a calling card for terrestrial intelligence. But more even than that it was a symbol that mankind stood at the threshold of its greatest transformation.

***

Biologically the human race had all but ceased to evolve. That, despite the popular, fanciful notion that people of the future would steadily acquire bigger brains and smaller, weaker bodies – eggheads whose limbs had atrophied and become useless through inactivity.

In fact, the human brain, in size and structure had reached its ultimate state of development perhaps 100,000 years earlier. A man or woman of the twentieth century, for all their cultured sophistication, had no more intellectual potential than their counterparts who painted on cave walls or stalked woolly mammoths in the ice ages of the late Pleistocene. Once there ceased to be an evolutionary premium of intelligence, the selective pressure to produce bigger brains had disappeared. And that had happened tens of thousands of years before, when humans began to dwell in semi-settled groups and cooperate and pool their skills. Then brain power, like muscle power. became a shared property, benefiting both the lowest and the highest intellects within a community. In consequence, not only the brightest individuals stood to pass on their genes. So, too, with equal likelihood, did those of lesser mentality. With mankind's mastery over its niche, albeit a global niche, its gene pool became effectively frozen, as that of the chimpanzee, the dolphin, and countless other species already had.

Yet there was one decisive difference. Before it ceased to evolve the mind of man had surged past some invisible barrier, had attained some critical mass or level of complexity. And as part of this development the brain had lost its ancient symmetry. Only within man's right hemisphere was the old, holistic mode of thinking perpetuated. The right brain continued to make no separation in nature, to invent no labels. Its perception, as it had always been, was of a timeless whole, a perpetual now, the universe in its simultaneous glory, unquestioned.

Not without reason would humans come to view other animals as being more at one with nature than themselves. In the mind of the cat, or of the eagle, or even of other primates there could be only the nebulous feeling of "is," consciousness without self-consciousness, observed without observer.

But for the early hominids that was not enough. These otherwise ill-equipped bipeds would have to become analysts and logicians to survive. They would have to begin to separate themselves, mentally, from the outside. To evolve further, to become truly human, the ape-men would have to construct a boundary between themselves and the rest of creation. And so, for the first time on Earth, there emerged a life-form that was conscious of itself as an individual, as an independent ego. And, further, a life-form that with the bright pencil beam of its newly developed self-awareness could isolate parts of the world – discrete objects and events – and then represent these by symbols, both within itself and without.

Thus did man the rationalist – the linguist, the scientist – awaken. Now there was a tiny part of the universe that called itself "I" and that with insatiable curiosity questioned the workings of everything around itself. Why did objects move, and how? What, ultimately, was matter? And time? And life? How far away the sun and stars? And what made them shine in the untouchable heights of the sky? Whence the universe itself? And where, in relation to all this awesome vastness, stood man?

***

Even as its first fragile craft broke ties with the sun, man conjectured what its descendants – should there be any – might be like in millions or billions of years to come. Genetic engineering and microelectronics and other such mushrooming fields already suggested staggering, almost limitless possibilities. Within a century after

Voyager's launch it could be that new animals – or humans – were being manufactured, gene by gene, to suit any purpose. Computers, self-replicating and with intellects ultrahigh, might be emerging as parallel, rival, or superior species. Who was to say that what once had been human intelligence would not soon be encapsulated, then hugely amplified – a blend of the biological and electronic? Without known precedent, Man hurled toward an uncertain future.

Maybe man would fuse with machine. Then living spacecraft, virtually immortal, might flit from star to star, forever learning and exploring. Or humans might more conventionally spread across the galaxy, restructuring planets, capturing sunlight with star-girdling rings or spheres as they went – yet still changing in disturbing ways. Adapting to local conditions, each colonizing group might develop its own "alien" culture, ethics, science, and art. Always there would be the chance of encounter with intelligent creatures. And who dared guess what that might bring? In time Homo sapiens could branch out along many strange and novel lines, altering much physically and intellectually.

A million years after stepping into space man might no longer recognize man – or what it had become. And yet still there could be a kinship, a warm feeling that here was a distant, half-remembered cousin. Even if someday all trace of humanity were erased, it might be reassuring to know that at least some sort of intelligence could live on. But, in Deep Time, was that possible?

There might be undreamed-of forms of life and mentality. With DNA alone the potential for variety was dizzying. And who could say what other types of matter might not also supply a suitable organic foundation? Yet, whatever its makeup, one thing it seemed life and intelligence had to have – a source of energy. Like stars.

While stars burned there was no fear of energy shortage. With billions of stars in each of billions of galaxies to choose from, life in all its manifestations could thrive.

But when all the stars had died, leaving only big black holes set against a backdrop of subatomic particles, what then? In such an unpromising environment how could even the most adaptable of sentient species fill its energy needs?

Perhaps by the Penrose mechanism — by ingeniously using black holes as garbage disposals. The method: An advanced civilization living close to a black hole would regularly fire projectiles at it. Each projectile would follow a carefully charted course to the fringe of the black hole's event horizon – the boundary from beyond which not even light could return. Here it would jettison a load of waste matter. Tumbling into the gravity vortex, the surplus material would be lost forever. Its rest mass energy, though, would be transferred to the parent projectile, plus an extra contribution from spin energy lost by the black hole. The projectile's gained energy could then be converted into a form suitable to supply the civilization's needs. Simple. How accommodating of twentieth-century man to find a solution to the energy crunch one million trillion trillion years or more in the future!

And yet, once all the black holes had boiled away, around AD 10^{100}, what then? With the entropy of the cosmos almost total, how could intelligence of any kind survive?

Consciousness or intelligence, it seemed, was linked ultimately to complexity. Sentient beings in Deep Time could continue only if there were some way to retain local pools of complexity or organization. And how to achieve that? Through lengthening periods of hibernation punctuated by bursts of energy-using activity? Or by escape altogether from that dismal future night, down a space-time tunnel, into a friendlier, parallel universe (should such exist) or a more energy-rich phase of this cosmos?

And what if the universe were not open but closed? Could life or mind endure as the walls of space-time drew in and in, until the cosmos, once again, was far smaller than a proton? Could it survive a Big Crunch and re-emerge in a new phase of the universe to continue its evolution? Might there even be archaic beings now, inside this cosmos, that had passed through previous cosmic cycles? Perhaps.

But if man and its technology were to progress unchecked even over a billion years, what would it really be like? Magical? Beyond perception or description? And if it were to survive for 10^{100} years? In the end even man's most farsighted prophecies were bound to prove hopelessly myopic. No more could a Neanderthal have foreseen holograms or Jupiter probes. No more could an infant have conceived its own adult thoughts.

* * *

For 18 billion years the particle had journeyed. Unconsciously. Unseen. In concert with all of the other particles that now made up the Earth, and the sun, and the beckoning stars beyond. Countless trillions of subatomic specks – dimensionless foci of energy – born in chaos,

147

scattered by the eruption of genesis, each making their own fabulous, convoluted way through space and time. Particles unaware of each other's presence. Yet, despite that unawareness, cooperating exquisitely still.

And so, aided by cosmic string, and by dark matter, and by its own innate propensity to come together, matter grew in complexity, by slow stages: nuclei, atoms, stars, planets. Organic broth. And then, in quickening succession, through higher forms of life.

There grew the first neural swellings that foreshadowed the brain. Consciousness and self-consciousness appeared on Earth – not in order that creatures might enjoy these extraordinary self-reflective powers but so that they should be better equipped for survival. And so, in time, came the monkeys, and the apes, and the first man-apes.

Almost overnight, revolution seemed to build upon dizzying revolution. First the lightning-fast physical change that took man's line from apelike Australopithecus in 3.5 million BC to modern Homo sapiens. Then the dramatic spread of culture from about 10,000 BC on. And finally the still more rampant growth of technology after the taming of steam.

Somehow bits of the universe had contrived to assemble themselves intricately upon the skin of an 8,000-mile-wide planet. They became thinking, self-aware organisms. And now these tiny ornate receptacles of stardust wanted to know about themselves, about the universe of which they were a part. Thus the universe scrutinized itself. And built instruments to study its farthest galaxies. And dispatched probes, made also of old star matter, to learn what they might. Probes like Voyager.

***

Did it see the irony? Did it see that in sending Voyager it was questing more deeply into itself – with an organized fragment of itself? Of course it saw. The words are written here. And now, at this moment, we share them.

Interconnections, always.

Nothing stands alone. That much, at least, twentieth-century man now dimly began to perceive. Every event in the universe was intimately linked to every other. So that, in truth, there really was only one, multiform, unfolding event – man. Or Voyager. Or the universe. Call it what you will. One event, indivisible.

And how could it have been otherwise? You are the Earth. And the sun. Interacting with each continuously. You are every particle in the universe, because every particle of the personal, the fictitious "you" communes by subatomic reactions continually with all matter and energy everywhere. You and I are everything. And only by the deceptive but all-crucial persuadings of the left brain were we ever convinced otherwise: that perhaps we ended at the boundary of our skin; that perhaps we were pathetically trivial motes of unusually complex matter clinging momentarily to a tiny rock in the vastness of space and time – with the cruel misfortune to be aware of it. And yet there is no beginning to you-I. No end, no true boundary. We are, in every way, the universe itself. When you-I think, the cosmos thinks. And when the cosmos thinks . . .

Already some had glimpsed part of the final truth, had felt it instinctively.

Some like Heindrich Hertz: "One cannot escape the feeling that these mathematical formulae have an independent existence and intelligence of their own, that they are wiser than we, wiser even than their discoverers, that we get more out of them than was originally put in."

Werner Heisenberg: "The common division of the world into subject and object, inner world and outer, body and soul, is no longer adequate."

Niels Bohr: "Physics tells us what we can know about the universe, not what it is."

Albert Einstein: "What interests me is whether God could have made the world in a different way; that is, whether the necessity of logical simplicity leaves any freedom at all."

John Wheeler: "Is the very mechanism for the universe to come into being meaningless or unworkable or both unless the universe is guaranteed to produce life, consciousness, and observership somewhere and for some little time in its history-to-be? The quantum principle shows that there is a sense in which what the observer will do in the future defines what happens in the past – even in a past so remote that life did not exist, and shows even more, that observership is a prerequisite for any useful version of reality."

Steven Weinberg, echoing Hertz a century later: "This is often the way in physics – our mistake is not that we take our theories too seriously, but that we do not take them seriously enough. It is always hard to realize that these numbers and equations we play with at our desks have something to do with the real world."

And, bridging the gulf between mysticism and science, Alan Watts: "Where do I begin and end in space? I have relations to the sun and air, which are just as vital parts of my existence as my heart. The movement in which I am a pattern began incalculable ages before the (conventionally isolated) event called birth, and will continue long after the event called death. Only words and conventions isolate us from the entirely indefinable something which is everything."

So, at last, the universe began to comprehend itself. Billions of years it had taken until there were minds such as these with the power to vaguely discern nature's scheme and their own remarkable place within it. Life's purpose, the origin and destiny of everything – with these tremendous issues the corporate mentality of man now strove. Why was it here? On this small ball of rock, circling an inconsequential star, in an average galaxy, in a macrocosm vast beyond deranged imagining?

***

Like all else, mankind had evolved, it realized now, with the talent to sense only what was needed for survival, to focus specifically on those aspects of the environment essential for staying alive. The senses, the brain, and the nervous system had developed to help discriminate between "safe" stimuli and survival-related ones. Which explained why no human (or other life-form) was ever consciously aware of each photon reaching it or of each subatomic particle it came into contact with. Even so, man – uniquely on Earth – would eventually learn of photons and electrons and the strange goings-on in the quantum world that were forever beyond conventional perception. For, through natural selection, man's brain had also evolved a left hemisphere that was analytical. Capable of symbolic reasoning, of verbal thinking. A left hemisphere that set up a boundary between its own contrived ego – its personal identity – and the rest of the universe. And that now wanted to know how the world around it was made.

To begin with, man's window on the cosmos was that furnished only by its own limited, biological senses. Through this narrow portal the early scientists glimpsed a universe ruled, apparently, by deterministic laws and in which humans were insignificant. The more the classicists learned of their cold clockwork cosmos, the more they themselves – as passive observers – seemed to shrink in stature and

150

relevance. Further, in the West at least, the dominance of left-brain thinking convinced men and women that they were just that – individuals, sealed off from their surroundings. The power of words and of labels by which humans conventionally identified themselves as independent creatures had blinded man to the truth. Thus did man begin to feel, like the word, static and detached, cut off from the real, frontierless world of nature. And thus, also, did humans convince themselves that they were small, discrete, organic forms, bounded by a membrane in space and by birth and death in time. Conventions! Mere social conventions were these, conjured up by the left brain. And yet they were so persuasive that people – those portions of the cosmos that called themselves "people" – spent their whole lives agonizing over their eventual death. Hungering for the perpetuity of a thing that never was. Failing to see that, in truth, they were smoothly flowing processes – no, a single flowing process – without definable beginning or end.

And yet, though the left brain beguiled with these conventions, it also was the means by which man eventually would break through to a greater understanding. As science – child of the rational mind – progressed, so it revealed new aspects of the universe that previously had been hidden to man's five "natural" senses. With penetrating instruments that its burgeoning knowledge helped to devise, man began to probe the strange, unfamiliar realm of the atom. Began to discover that, on levels inaccessible to normal perceptions, reality behaved in ways that defied common sense. And so, in time, man learned (incredulously at first) of the bond between observership – its own role as self-aware observer – and the universe as a whole. Nothing existed "out there" until it had been consciously observed.

Even as Voyager, with the particle hero aboard, began its endless flight, man's mind grappled with this extraordinary new thought. That there had to be self-awareness, human or otherwise, to summon reality, this reality, into being.

And, at the same time, other pieces of the great puzzle began to fall more neatly into place. For it seemed also that the universe was fantastically, almost unbelievably, contrived. Only minute differences in the laws of nature or in the values of the fundamental constants would have resulted in a cosmos in which intelligent life of any kind could not have evolved. Had the Big Bang been slightly less violent than it was, then space-time would have collapsed before stars could have formed. A more vigorous blast and matter would have dispersed

151

too quickly for any organized structures to coalesce. Then, too, if space had emerged from genesis with fewer than three dimensions, nothing as complex or interconnected as a human brain could ever have evolved. While in a space of four dimensions or more, the orbits of planets around stars would have been unstable. The requirements for life – intelligent life in particular – were many and specific. And the universe it transpired, against all the odds, satisfied every one.

Now, finally, man began to see why.

Observership, quantum science showed, was a prerequisite for the selection of reality. Reality, cosmology showed, was finely tuned so as to inevitably spawn observership. Necessarily, imperatively, the two went together, for only in this way could there be a universe at all.

And yet, to begin with at least, it was too much for many to grasp. Men and women, caught up in the treadmill of daily life, paid little heed to the gradually dawning solution to everything. Why should they when there were so many more immediate thoughts to occupy them? So many apparently larger issues like bank accounts and shopping trips and job prospects and pension plans? Compared with these, of what relevance was genesis, or the evolution of the universe, or the relationship of mind to matter?

But, in time, awareness would come. And then man would turn away forever from the womb of parochialism and the trivial, transient cares of individuals and communities and nations – conventions all without substance. And would look with new, clearer vision, both inward and outward, at the universe of which it was the essential, cumulative product.

***

How much like a human being the cosmos as whole had developed. And by the same token how true had proved the suspicion of the mystics, with their primal, right-brain awareness, that man was a microcosm of nature. Both infant human and infant cosmos began apparently from states simple and symmetric, from a featureless genesis egg offering no clue as to what it might become. From early oblivion both progressed steadily, in accord with some hidden, inner code, to consciousness, waking up gradually to the tremendous fact of their own existence. And then questioning: Why were they here and how had they been created? Why, against every expectation, could they understand the laws that governed their smallest action? What was their purpose? And where were they bound?

152

"As above, so below," the mystic would conclude. And there was much truth in that. For by turning within, the mystic read the self-same laws that operated in the world of physics, though from a different perspective. Whereas science surveyed the microcosm outside the seeker, mysticism gazed within at the seeker itself, until the seeker merged with the sought. Science was outer empiricism, province of the left brain; mysticism strove instead, by quieting the left brain in meditation, to directly experience nature with the right. And now, even as Voyager took to the boundless skies, it seemed that these two very different approaches were converging fast upon their common ancient goal.

"Science cannot solve the ultimate mystery of nature," proclaimed Max Planck, one of the twentieth-century fathers of quantum mechanics. "And that is because in the last analysis, we ourselves are part of nature, and, therefore, part of the mystery we are trying to solve."

As Voyager plunged into the ebony night beyond Neptune, particle physicists and cosmologists on Earth consummated an unexpected alliance out of which was born the Theory of Everything. For a long time science had sought it, the Physicist's Stone, the all-encompassing set of equations that would unify every physical phenomenon in nature. That would embrace every subatomic particle and every force within a single, stunningly beautiful, mathematical model.

And now the last piece of the cosmic puzzle slipped quietly, almost unnoticed, into place. For as Voyager left the solar system, bound for the high interstellar seas, scientists finally answered the question that Einstein and others had posed years before. Yes, there was indeed, beyond doubt now, only one possible mathematical description of the universe. Only one description that provided for the eventual emergence of intelligent life and yet was self-consistent in all its detail. The Theory of Everything.

***

Time passed. And mankind, driven as ever by its need to explore, pursued its robot craft to the stars. To Alpha Centauri and Tau Ceti and Ross 248 and a score of other suns in the solar neighborhood. And then, in ever faster ships, miles in length, men and women ventured, to build cities on worlds unimagined. And then, from these remote outposts, other expeditions set out in all directions to penetrate still deeper into unknown tracts of the galaxy – and beyond.

A million years had passed since Sputnik was hurled into Earth orbit. A million years in which Voyager had travelled forty light-years from home. And in which its creators had been transformed past all recognition. No longer was man identifiably a single species. Or a species at all. No longer was it constrained by its old organic body and brain. Those long eons of patient biological evolution on Earth, it was clear now, had been merely the precursor, the preparation, for an incomparably more dramatic phase of technological growth. The boundary between human and human-made was no longer discernible. What had been computers were now continuous extensions of the brain. What had been spacecraft were now adjuncts to the living forms of creatures who flitted from star to star, even from galaxy to galaxy, like butterflies among flowers.

For now man more fully understood its cosmic role. And in its growing enlightenment it had shrugged off its pupal shell – its earthly incubus and limited human form – and reached outward to the greater universe beyond. The universe which it alone had created and which, in turn, had created it. The universe which, simultaneously, it was, and of which it was the most essential part.

Now, with all aspects of the Theory of Everything at its disposal, man could aspire to technologies that were truly godlike. Whole stars and planetary systems it dismantled and toyed with and reassembled at will. Entire galaxies and groups of galaxies, in time, it won control over and landscaped at its pleasure. And yet such stupendous feats of cosmic engineering were but a preface. For eventually this thing that the human race had become gained mastery over space and time itself, so that it could communicate or travel instantaneously over any distance. Could marshal even the energy of the vacuum and the residual blast of the Big Bang for its disquietingly grand schemes.

**

For more than a hundred billion years Voyager had endured. While space-time expanded and the galaxies aged. And the star-making machinery of the Milky Way . . .

Was redesigned.

Now bright pulsating rivers of energy coursed this way and that across the Milky Way, purposefully supplying the needs of a galaxy with one mind. A galactic being that already had begun to absorb and assimilate other star cities around it and that in a mesmerizing blend encompassed as-yet-unconscious matter, intelligence, and hyper-advanced technology. That embraced all of what once had been humanity. But that was now far, far greater – and becoming greater still.

Ages ago the metamorphosis had begun – the melding of individual human minds (together with those of other sentient beings) into a single consciousness. Through brain-computer links and knowledge repositories, man had come to share its most intimate thoughts, to reconcile even the smallest differences in opinion. So that eventually whether there were many separate nodes of awareness or just one collective mind was hard to discern. In the end, indeed, it made no difference. Man had long since broken free of the shackles of petty labels and conventions. If it so wished, it could still conceive of itself as an individual, and so enjoy the privilege and private identity of the observer. And, in so doing, it retained the focused, rational powers of the left brain, but now awesomely enlarged. At the same time, though, it was aware, without a doubt, through right-brain perception, that it was the single, undivided universe, still evolving toward a state that even now it could not fully comprehend.

Of this much it was certain. There were not many possible forms that the cosmos could have taken. In the final reckoning there was only one. Of all the feasible mathematical descriptions of reality, only one was self-consistent, self-sustaining. Only one could make the essential jump from potentiality to actuality – the one that could spawn intelligent observership. And why? Because only through observership could the universe become, by progressive stages, conscious of itself, comprehending of itself, and, ultimately, selective and creative of itself. Only as mind emerged could fire be breathed into the one global mathematical model of reality that could yield mind! As mind matured and, by intimate association of aware observership and mathematical theory, came to understand rationally the cosmos of which it was a part, so it began to draw the cosmos into being. Presently it astonished itself by realizing that it had this power – that it was the self-creating universe. That it was inevitably so.

***

Thus was the terrific conception revealed. That mind was what transformed the abstract mathematical relationships of the world into perceived reality, gave substance to the equations. That there was nothing concrete outside the mind. That mind created reality – inner and outer – and, at the same time, allowed the underlying mathematical relationships, to be symbolized and better understood.

Of course. It seemed so plain, so self-evident to man now. The rules governing the universe could only be made manifest – be expressed and given tangible essence – by the coming together of a real cosmos with real space, time, energy, matter, and forces. The play demanded the players. By the same token, the physical universe was delivered to the threshold of reality by those rules – just as the pieces in chess are made potential by the rules of the game. The final crossing of the threshold came, as quantum science had conclusively shown, with intelligent observation. Thus did mind, matter, and mathematics each subtend and make possible the others.

And genesis? That was never in truth the source of the cosmic stream. For the stream began, not on the ground, but as rain from the sky, which came from the sea, which was fed by the stream. Which is to say that genesis – the Big Bang – was not the beginning but merely the point from which the universe had seemed to come. Matter and energy and the rules governing them could spring into existence only through the medium of mind. So that before genesis came mind. And before mind came genesis. Both joined in a closed cycle of time stretching from past to future and back again.

And of course for man there had to be a future. Of that the minded galaxy was absolutely sure. For the only way in which it was being sustained and made real now was through observership by some yet greater intelligence to come. Awesomely powerful though man had become, it was still far, it knew, from being fully aware. Still it had not yet penetrated back to see the finest nuances of the Big Bang, nor did it even understand in every detail how its own thought processes worked. There was much yet to learn, much of reality yet to observe – and hence be created. So that man's present existence demanded that intelligence continue to exist. And more, that this intelligence, this self-awareness, continue to spread throughout space until it had gained control over all matter and all forces. Ultimately, every part of the universe had to evolve to sentience, to become a single, eternal mind.

***

This was the covenant of nature. And it had been realized, however vaguely, long, long before man inherited the galaxy. Before the Theory of Everything had been fully revealed, before humans had claimed Mars or Ganymede for their own, before even Voyager had felt the gravitic lure of Prospero, the astounding conclusion of quantum philosophy had begun to filter through. Man and mind had to continue to exist and grow for all time. There could be no end to intelligence once it had come into being. Even now, at the close of the twentieth century, we sense it. You and I are the infant cosmos, still only dimly aware, still only conscious of things immediately around the reality generators that are our minds. We perceive only dully, over a small range of wavelengths of light and sound, and we comprehend structure over only a narrow range in space and time. But, eventually. we will see X rays and gamma rays, radio waves and gravitational waves. And subatomic particles. And whole galaxies in their most intimate detail. We shall see and understand all that there is to know. What we are today will evolve to become a single universe-wide mind, so that every particle in space will be within this cosmic consciousness – free, but aware. Every particle of which you-I are made will ultimately be reconstituted in this universal mind, along with everything else. Given such a prospect, we need hardly fear our own personal deaths. For nothing ever dies. And in Deep Time we shall be as one.

And so to the resolution, the stunning climax of it all. The answer to the mystery of life and being. The Fugue in C. In C Major. Or in creation. A Bach fugue. How very appropriate. Bach, the most mathematical of composers and the most subtle. Lover of hidden meaning, of theme and variation, of endless cycles.

And how very appropriate, too, that this Fugue in C should rise up through the scale until it returned, surprisingly, to its starting note – but now shifted up by one octave. Like the universe itself which began on the note of creation. Then progressed through various ascending stages, until it became conscious. And then self-conscious. And, finally, self-creating. So that, in Deep Time, it looped back to its origin again, and yet along the way had elevated itself somehow, from chaos to cosmic consciousness – evolved to the awesome stage where it could create itself.

Still it was hard to comprehend, to accept. If the universe could only be created in its own future, then how could it ever have had a past? Surely there had to have been some special point of origin? But no. What was needed was a more panoramic view in which the universe, past, present, and future, was seen as having always been there – a permanent, all-encompassing space-time entity.

Of course it was natural for man, whose left-brain consciousness produced the illusion of "passing" time to think of past and future as somehow being different in status. To dwell, moreover, on that elusive moment called now which transformed the potentiality of future events into the actuality of the past. But "now" was, in truth, only a chimera. Every point in space and time coexisted with equal importance. The future was there from the beginning as surely as was the past. So that the means to reach back in time to summon the cosmos into being was ever present.

***

Across the farthest intergalactic gulfs man's maturing mind spread. Annexing more of the universe. Bathing previously dull corners of space with the searing light of its intelligence. Now each galaxy was like the neuron of a cosmos-filling brain, while the voids between galaxies were latticed with the pulsing currents of inconceivable thoughts and ideas.

What once had been Voyager was now incorporated within this superconsciousness. The spacecraft's subatomic frame, long since dispersed, had become part of the very means by which cosmic man thought. Everything that had been was included in the universal mind: the particles of the ancient sun and Earth. And of the reader. And of the ink and paper of these pages in which such prophecies were made.

In Deep Time we shall be as one.

***

What once had been the particle was now the whole. Had it ever been otherwise? In the universe there was no – could be no – isolated parts. Only a whole. And now this holistic thing that was everything looked in upon itself. Saw that it had evolved to become, at every point, a living, sentient organism, a cosmos-wide being. Mind had become one with all of nature. So that no longer was mind peering out on the rest of the universe and mapping it and trying to understand it symbolically. The physical reality had become coincident with its own symbolic mapping. The thoughts of this all-embracing cosmic mind were no longer an abstraction of reality, they were reality. So that when the cosmos thought, its thoughts sprang spontaneously into physical being. At last mathematics and physics, God and man had become one.

***

And now God surveyed all of space and time, all of matter and energy, all that it once had been and was now. God, man, the universe –

158

this awakened mind – strove to gather into its consciousness every remaining fragment of the wealth of existence. Strove to perceive every event at the minutest level so that reality, everywhere, everywhen, would be focused to crisp perfection by retroactive causation.

And then God turned to its ultimate project. To create within itself the genesis event – the Big Bang.

***

There were no landmarks. No galaxies. No stars. No elaborations of matter of any kind.

Yet somewhere across the stupendous ravine of Deep Time that lay ahead there was a mind. With piercing, all-seeing intelligence it gazed at a single point in space and time – in no way different from any other – and thought. Because it already existed in the time-to-come, it had to be created. And since it was all there would ever be, it had, inevitably, to create itself. Simply by being and by observing, it summoned up its own existence, its own embryonic reality. A reality whose rules permitted creation by observation and that was so contrived as to lead necessarily to observership. A reality that had to be underpinned by one very special set of rules – that unique set which, 15 billion years later, humans would begin to decipher as the so-called Theory of Everything.

Out of nothingness there arose a stir – an eddy, a flicker, a something inconceivably small. And the mind of man watched. And saw that it was good. That genesis, once again, was under way. And that presently, within itself, man would bear new havens, new stars and planets, a sun, and an Earth under whose blue skies men and women would renew their quest for Deep Time.